THE
TOOLS
THAT BUILT
AMERICA

Other books by Alex W. Bealer

THE ART OF BLACKSMITHING
OLD WAYS OF WORKING WOOD
ONLY THE NAMES REMAIN: THE CHEROKEES AND
THE TRAIL OF TEARS (JUVENILE)
THE SUCCESSFUL CRAFTSMAN: MAKING YOUR CRAFT
YOUR BUSINESS

THE
TOOLS
THAT BUILT
AMERICA

ALEX W. BEALER

Illustrations of tools by the author
Photographs by Dr. John O. Ellis

BARRE PUBLISHING
Barre, Massachusetts
Distributed by Crown Publishers, Inc.
New York

Library of Congress Cataloging in Publication Data

Bealer, Alex W.
The tools that built America.

Bibliography: p.
Includes index.
1. Woodworking tools—History. I. Title.
TT186.B3   1976   684'.082'0973   75-45424
ISBN 0-517-52405-8

To my eldest,
Janet Bailey,
and her husband, Ivan,
who are still building America
with their craftsmanship

A SPECIAL ACKNOWLEDGMENT

Much of the value of this book lies in its photographs. These provide a certain reality to the text and the drawings. Without them the story could only be half told.

For old tools are not mere inanimate objects or historical curiosities. Each, when it was used to produce beauty, gave itself and the craftsman who used it a nameless immortality expressed through the grace of a molding, the strength of a tenon, the precise dimension of a carving.

My deepest gratitude and respect, then, to the photographer, Dr. John O. Ellis. Dr. Ellis is the perfect working companion for me. We have worked together on many projects of common interests since boyhood. We understand each other to the point of being almost one in our joint endeavors, including this work.

Dr. Ellis's photography has made this book whole.

Further thanks is due to Dr. Frank Allbright, the venerable and enthusiastic archaeologist at Old Salem in Winston-Salem, North Carolina. Dr. Allbright's unstinted efforts led us to many treasures that demonstrated the functions of old hand tools, including that rarest of archaeological finds, a pit-sawn joist in the roof of the Brothers House.

To him and to Old Salem, many thanks.

CONTENTS

THE
TOOLS
THAT BUILT
AMERICA

INTRODUCTION

The tools that built America are truly wondrous. The early hand tools—simple yet ingenious, beautiful in form and substance yet efficient—are the most interesting. As varied as human nature, each hand tool could almost be endowed with a life of its own through the skill and love and imagination of a craftsman. Each was a medium of expression, much like an artist's brush and paint or a sculptor's chisel.

This book is limited to the woodworking tools used by the frontiersman, the carpenter, housewright, and the cabinet-maker. It regretfully passes by the rich variety of tools used by masons and silversmiths, blacksmiths and wheelwrights, tinsmiths and white coopers, and many other special tradesmen of Colonial times. These craftsmen, too, were essential, directly or indirectly, for building and furnishing houses, but in this book their tools and methods would distract from the basic contributions of carpenter and cabinetmaker.

Of course, the tools that built America were shaped by the needs and conditions of America, from the first settlement at St. Augustine in the sixteenth century until the present. The basic needs of the first European Americans were the same as the needs of Americans three and four hundred years later. In addition to food and clothing, both of which were expendable and easily replaceable, they needed shelter—a different matter altogether.

Shelter was permanent, or relatively so. Homeplaces, in Europe and America, consisted usually of a colorful variety of buildings unknown to most modern Americans. Even the dwellings found in crowded towns, which followed the pattern of medieval Europe, often consisted of a house and several outbuildings.

Each family had to have living space and sleeping space and cooking space, all under shelter. Sanitary facilities, such as they were, often required a separate building. Storage space for tools and gardening equipment was needed, as well as protection for the water supply, and sometimes a smokehouse and a cooling house were part of a town dweller's establishment. Horses were the means of transportation by land and in addition many town house owners kept a cow. These domestic animals also required shelter, and a place was needed in which to store animal feed. Often separate houses were built for servants, in both town and country. Altogether the basic need for shelter was intriguing in its variety.

While food and clothing, and furniture to some extent, could be imported from the mother country, shelter could not. The concept of prefabricated houses was unheard of in early Colonial times; there was no adequate transportation to deliver a house across the wild and treacherous Atlantic to the shores of the New World. Houses had to be built on the site from materials found nearby. The only exception known is the agreement by which Colonel William Bull of Charleston shipped prefabricated houses to the new capital of Georgia, Savannah, in 1733.

Of course, the outstanding phenomenon of the western shore of the Atlantic was the Great American Forest. Here was a woodland of vastness and variety that had not been known in Europe since prehistoric times, except perhaps in Russia and Scandinavia. It offered limitless potential for all things made of wood: wagons, casks, houses, furniture, tools, ships, charcoal, fences, weapons. Granted, there were certain areas in the barren western desert settled by Spaniards that offered little wood for building; wood was used only to supplement building by adobe blocks there. For the most part, however, European colonization of North America, especially the settlements of northern nations, occurred in places where wood was the most plentiful resource available.

As a consequence, most of the houses and accoutrements of Colonial America were built of wood, frequently wholly of wood, even to the exclusion of iron nails and hinges and latches. There were exceptions, certainly, but wood was the prevailing material.

This followed ancient traditions of building in Europe. There too, wood, easily worked and replaceable through growth, had always been favored as the major building material for dwellings until defense needs and size (of great castles and cathedrals) required the strength and massiveness of stone and brick. Indeed, brick was developed in the lowland countries, those relatively treeless portions of coastal Europe. But the techniques of brickmaking and brick masonry were not exported from Holland and Flanders until around the fourteenth century and brick became a substitute for wood only when the forests of France and England and Germany were first beginning to disappear noticeably.

There was plenty of wood in Colonial America, but power, as compared with the sources of power in modern times, was greatly lacking. Until the 1770s, shortly before the end of British control of the colonies in America, when James Watt conceived the steam engine, available power was limited to simple inventions involving the wheel and the harnessing of wind, water, and animal power.

There was animal power, generated by horses, oxen, and a few dogs, but this was applied mainly to transportation and farming, and only the simplest of machines could be operated by animals. Wind power was used in western Europe and her colonies to run sophisticated mills which ground grain into flour and meal, and perhaps for a few additional tasks such as running grindstones and wood lathes. Wind, of course, was also used to propel the ships of the world from one continent to another, for purposes of war and travel and commerce.

Waterpower, established since Roman times, offered the apogee of automation in all aspects of Colonial life. The water mills of the day were marvels of mechanical ingenuity. They could be designed to grind corn, operate the great bellows of an iron furnace, actuate the trip hammers of an industrial smithy, or run a sawmill.

Water mills were also used to operate turning lathes and jigsaws, and to provide auxiliary power for pulling large-sized molding planes. Sash saws which converted logs into lumber, one or two or three boards at a time, were operated by water. Sash sawing was a slow and ponderous process which appeared fairly late in history, at a time when the quantity and quality of steel were finally available for making large saw blades; it disappeared generally around 1850 when the circular saw blade was universally adopted.

Water mills are interesting to the antiquarian because man, early in his history, invented them to utilize that elemental source of power, weight. Weight, rather than mere size, was an important factor in all of man's tools, mechanical and otherwise, until the steam engine was developed late in the eighteenth century. Water, of course, in inexhaustible supply in most regions, provided about eight pounds of weight per gallon. In regions above the fall line, water was mainly concentrated in streams where its weight and gravity drew it downhill inexorably to the sea. Man utilized this combination to turn his wheels and provide his power. Fortunate was the wheelwright and turner who had a small stream near his shop to turn his lathe.

Most of the power used by the woodworkers of Colonial times, and before, was that ubiquitous, now waning form of energy known colloquially as "elbow grease"—the muscles of man himself. For countless millennia it was the power of a single man, or groups of individuals working together, that built houses and made furniture and did all the other labor required by refined people who had learned to live comfortably despite a hostile environment.

Manpower offered many advantages and was, indeed, absolutely essential in premachine days. First of all, it was portable. Houses and other buildings, and even cabinetmaker's shops, could not always be built next to streams, nor was it worth the effort to build a dam and sluice and waterwheel and machinery just to build a house or shop. A craftsman could always travel, with his talents and his tools, to the site of a planned house. Heavy machinery could not.

The facilities of transportation were limited to horse-drawn wagons and boats propelled by horses, oars, poles, or wind along the shores of oceans or on rivers and lakes. Transporting faraway material for a house to the building site was costly and time-consuming and, again, wasteful of effort. It was easier to use material found in close proximity to the building site, and in most instances the preparatory work needed to transform trees into boards and boards into molding was done on the site, by manpower.

Any modern man who looks at an old house made of wood, such as the seventeenth-century Fairbanks house in Massachusetts, and speculates on how it may be duplicated with manpower and hand tools, is filled with a sense of inadequacy. How, for instance, could all those clapboards be made by hand without taking years of a man's life? How could the countless shingles on the roof be made without the facilities of a huge modern sawmill? How long did it take to cut and square all the timbers in the house frame, all the posts and studs, plates and joists and braces? Who could possibly have the power or the patience or the skill, or for that matter, the time to form all the mortises and tenons, the panels and delicate moldings in a house the size of the Fairbanks house? Who would have the brute strength to assemble the frame and raise it into position? The Fairbanks house and countless other houses and barns in America and Europe, some standing for eight hundred years, are evidence enough that somebody did, but how?

Part of the answer lies in the attitudes of men in past ages, part in the tools man had developed over thousands of years and the concomitant skills to use these tools. Attitude in combination with tools provided a simple but admirable technology that satisfied the needs of the times.

In early times in America, and in preceding eras all over the world, man simply did what he had to do with what he had available. Patience was a necessity for him because he did not have access to the sources of power his modern descendants now enjoy. He followed the philosophy expressed so simply, and so frequently, by the Chinese proverb: "A Journey of a thousand miles starts with a single step." He expected his fabrications to take time. He finished them bit by weary bit, but he had a fine sense of accomplishment when the job was done. To make his job easier and to increase his production he invented tools—marvelous, ingenious tools.

Volume, however, was rarely the primary objective of a workman; quality was, and beauty was the second. We may be certain that not all Colonial workmen achieved either goal, but those who did left a heritage of taste and high standards that inspire us today.

Of course, societal conditions in Colonial times were favorable toward the attitude and patient methodology of the handcraftsman. Back in the seventeenth and eighteenth centuries the population grew slowly. Once a man settled in a spot it was likely that he would stay there until he died and that some of his descendants would remain in his house, using his barn and outbuildings and furniture until they died and their children took ownership of the handcrafted property. In those slow days, particularly in America, land was not nearly so valuable as buildings and furniture, so a man's house was

6

often the major item of his estate. So valuable were beds and certain chairs that they were considered prized legacies, as attested by many an eighteenth-century will.

The point is that although houses and pieces of furniture took a long time to build, they were built to last for generations. The labor that went into them was painstaking. The better houses and furniture became monuments, sometimes to an individual and always to the race of man.

Almost everything was provided through the use of hand tools, which were, of course, operated by manpower, shaping timbers or fine moldings by removing one chip or one tissue-thin shaving after another until the desired shape was made. Eighteenth-century hand tools showed some refinement generally over the tools used by craftsmen in the Dark Ages, but not much, and some tools of Roman craftsmen, which preceded the Dark Ages, were often more highly sophisticated and efficient. In most cases the tools of the eighteenth century were no different in principle from the tools that might have been found in the carpenter shop of Joseph of Nazareth. Indeed, certain of the basic tools of Colonial America, such as ax and adz, were prehistoric in origin.

By Colonial times the ax, adz, and hammer, in all variations, had been made of iron and steel for five or six thousand years. When Europeans first came to America they used axes and adzes of the same basic design found in prehistoric axes all over Europe and Asia. The hammers they brought with them were identical in principle to unearthed claw hammers used by the pre-Christian carpenters of the Roman Empire.

Some of the relatively recent refinements in the woodworking tools found in the chests and shops of the first American craftsmen in wood included: sophisticated planes, chisels, and augers of many different shapes and sizes; a variety of saws and hatchets; squares, levels, and scribers; screw-cutting tools and dowel planes; a number of special variations of all these such as twibils, holzaxtes, and commanders (rammers). Prototypes of fascinating variety, however, so lacking in modern tools, could have been found in the kits of Viking longship builders, and medieval cathedral carpenters. They had, almost all of them, been used by the craftsmen who made and fitted the panels of seventh-century Middle Eastern mosques, or Hindu palaces in the age of Alexandria, or by the furniture makers of Homer's Greece. The hordes of woodworkers on Solomon's Temple would have immediately understood the tools of early Colonial carpenters, and vice versa. The Colonial tools were a bit more versatile and refined, but there was virtually no difference in principles or types.

Really, there was no reason for any differences. The wood which was worked in early Virginia had the same qualities as the wood which grew in Paleolithic times. The man who lived in the England of Elizabeth I, or the Portugal of Henry the Navigator was the same man, physically and mentally, who served Solomon or manned the ships of Ulysses. Conditions had changed in the social and cultural sense, but attitudes toward life had changed hardly at all. Tools in seventeenth-century America were still extensions of the man, just as they had been in the seventeenth century B.C. The dependency of man on his environment was the same.

In America this relationship was intensified.

When Europeans first began to settle the forested wilderness of the New World, largely in the seventeenth century, they faced new conditions for which there was no available accumulation of skill or distributive and mercantile techniques, and certainly no established technical community. Shelter for man and beast, for example, had to be built quickly because there were no relatives or friends with whom a family might stay while shelter was being built. These conditions continued for almost three hundred years until the seemingly inexhaustible American frontier finally ran out. As a consequence the American attitude was altered. Efficiency was needed as well as skill. Time and winter winds waited for no man.

One consequence of these conditions was the development of the distinctive American ax. Early settlers in America brought with them the ancient European felling ax, a mere strip of iron shaped around a mandrel to form the eye, with a steel bit welded into the end of the long, narrow blade. It had no poll, and it had no balance. Such a tool was adequate for its uses in Europe, but its poor balance was a hindrance to the efficiency of American axmen who needed to build cabins and houses in a hurry.

The American made the eye flatter and longer to stabilize the wooden handle. He added a poll behind the eye and he shortened the blade. The result was a well-balanced tool with enough weight behind the edge to give it a bite merely through the momentum of the swing. Little effort was required to swing the ax, and each swing cut deeper than its old-fashioned predecessor had done. With it the American woodsman was able to conquer and use the supposedly limitless forests of America. Later, the American ax was adopted by woodcutters of most civilized countries of the world, although certain nations, such as Portugal and its former colony, Brazil, continued to use the inefficient felling ax into the 1970s.

Toward the end of the frontier period, in the 1850s, American lumbermen made still another contribution to ax design when the double-bit ax was developed. Some of the Scandinavian countries had used the double-bit ax for generations, but the Americans perfected it and the thin, perfectly balanced double-bit ax was extremely important in providing the vast amount of lumber needed to build America's cities and farms during the period of the nation's greatest expansion.

Both of these American designs were developments of one of the earliest tools made by man, and under environmental conditions similar to, if not exactly like, the forests of prehistoric Europe; the attitudes, however, of the early European and the early American, separated as they were by thousands of years of cultural evolution, were entirely different. The early American attacked a primitive environment with a rich cultural background behind him, holding established standards in housing and furniture and ways of living. He had no desire to revert to living in rude tents or caves; he wanted the comforts of house and home inherited from his ancestors of a thousand years or more. He expected the basic conveniences found in the houses of England, Holland, Sweden, Spain, and

France. Without the established economic amenities of specialized crafts and trades, without transportation facilities as a system of distribution, the American perforce had to provide himself with what he needed. The old-fashioned, ill-balanced ax simply was not an adequate tool to give him the materials he wanted in the time available. He therefore developed a tool that could satisfy his wants more quickly. The American ax was a fourth dimension, as it were, in the settlement of North America.

Other elements of cause and effect and American conditions of environment and population brought on additional developments in woodworking. For instance, the cut nail was first introduced in the New World about 1790, following on the heels of the decisive victory at Yorktown. The Americans used the ancient principle of applying weight, through a water mill, to supply nails plentifully and cheaply to builders. In that day of hand-hewn timbers and mortised joints the introduction of the cut nail had only a mild effect on the economy of the day. In fifty years, however, it became a major factor in American building and was soon adopted by the entire Western world.

There were other conditions, however, that prevented the cut nail from being adopted as quickly as it might have been. One was the sheer human inertia; another the absence of communication, both in transportation and in advertising; still another was the lack of capital, for nails cost money and hard money was scarce.

Before 1790, only rich men living in towns or on great plantations, whose houses were covered with clapboards nailed to the frame with (often imported) wrought nails, could afford nails of any kind in quantity. All the rest, the majority of farmers and backwoodsmen, stuck to the ubiquitous log cabin, the small saltbox, the stone house, or the house built with walls of solid brick made on the site.

The invention of cut nails was probably the first important step in America's Industrial Revolution. But in the next fifty years in Europe and America there would be other industrial developments, which would affect, directly or indirectly, the means of manufacture and the types of houses and furniture of Americans and, eventually, the world.

In eighteenth-century England, for instance, James Watt had made and operated a primitive steam engine; the mechanical brilliance of eighteenth-century France had produced a milling machine, the principles of which would later be applied to woodworking machinery; in America that genius of mass production, Eli Whitney, was hovering in the wings. Whitney, of course, was concerned with the cotton gin and the mass manufacture of firearms rather than woodworking—Whitney's inventions, however, particularly the cotton gin, affected the growth of the new country and created new needs for houses and furniture.

Even the fine cabinetmakers of the eighteenth century adapted to the demand born of the period's growing affluence. Diderot's *Encyclopedia* illustrates quite large cabinetmaking establishments in France with specialized workers attending to the various aspects of furniture making, prototypes of the assembly line workers of a century and a half later. In England the king of cabinetmakers, Chippendale himself, operated a virtual factory in which he supplied the designs, supervised the work, and maintained a remarkable volume of production. Goddard in Newport operated similar factories. The organizational concepts of these eminent artists were to form the basis of the Grand Rapids school of furniture, which appeared a couple of generations later. Unfortunately the latter-day cabinetmaker was all too often more interested in production and organization than art, and suffered the artistic disadvantages of vast power to run the machines that replaced the skilled journeymen craftsmen of Goddard and Chippendale.

And then there was Ben Franklin, who one rainy day caught lightning with a silken kite. The bolt which ran that day from kite to key had in it such a potential of creature comforts and power that it would in time destroy the economic importance of the handcraftsman in wood and relegate his position to history within a mere two centuries.

It took many years for all these sometimes unrelated developments to virtually eliminate the village craftsmen and literally effect the disappearance of the usually anonymous journeyman cabinetmaker who hauled his tools and lumber from plantation to plantation in a wagon to bring a touch of refinement to isolated baronies of the Old South. Yet these ideas were all to be important influences on the life and society of generations to come.

For the forces of history cannot be sifted and separated and made to apply only to specialized fragments of man's progress. Each force affects the others, and is affected by the others. However unrelated these industrial developments were at the time of conception, all affected the future tools and techniques of the woodworker, and the design of what he was to make.

About 1840 the carpenter's life was changed with the growing need for new houses and the consequent introduction of "balloon," or frame, construction for all sorts of buildings. Using this technique, which consisted of building a house frame of two-by-four timber which was then covered with clapboards on the outside and sometimes ceiled with boards or plaster on the inside, carpenters learned to depend on hammer and handsaw and began to discard the ancient adz and twibil and a host of other hand tools which had been used throughout antiquity for the careful mortising and joining of house timbers. The cut nail industry was well established by 1840 and its cheap production, distributed by relatively dependable steam-driven vehicles such as railroad trains and riverboats, nurtured the new building techniques. In the same general period the circular saw appeared. This efficient device could be run by semiportable steam engines and was not dependent on a waterwheel for power. It supplied all the two-by-fours and clapboards a growing country could use.

And so the houses of America began to change in form, in size, and in durability. If a rich merchant or transportation magnate wanted a huge house without waiting a couple of years to have each sill, post, and joist squared with a broadax, and each clapboard and shingle riven by hand, then he could have it in a few months in the new industrial society aborning in the 1840s and '50s. It was quick and easy to order all the lumber from a mill, have the stock delivered by train or steamboat, and then have local carpenters nail the structure together.

The skills acquired in using adzes and broadaxes in carpentry were transferred to supplying crossties for the railroad tracks just beginning to be laid in the 1840s. The carpenter, however, was left with his planes for another generation. Much of the lumber received from the sawmills was rough, and the carpenter smoothed it with a hand plane until the 1860s, and in some country areas until World War I. Also he kept his sets of molding planes, his tongue-and-groove planes, door mold planes and others until the mills, with readily available steam engines, set up complete planing mills that supplied a variety of molding for inside and outside finish as well as finished boards.

Modern types of molding planes, however, with steel frames and interchangeable bits continued to be sold by Sears, Roebuck and many hardware stores until the 1940s and 1950s. But after World War II these planes practically disappeared, replaced by a plethora of home workshop power tools of hand and bench variety.

By the early 1900s about the only use a carpenter had for a plane was to trim doors, and he seldom carried with him more than a jack plane and a smoothing plane. By the 1940s motorized hand planes, with rotary bits, were available and were used by the craftsmen who finished banisters in large buildings, or who specialized in hanging doors. Such power planes destroyed the ancient, satisfying rhythm which had become part of the woodworker's skill, but they also increased production and income, always important factors to a professional.

It was the big mills, though, that made the plane a secondary tool in the carpenter's toolbox. Small electrical bench saws replaced the magnificent sets of planes used by the old-time cabinetmakers, and mills provided the modern housewright with almost unlimited production of ready-made molding and panels. But the old variety, the individuality of the trim of older, handmade houses, was of necessity limited in mass-produced moldings. Again, this limitation was a matter of economics. Rotary bits, or cutters, could be made to any design, but the process was expensive. Thus the variety of cutters was limited. And, too, changing bits required skilled labor, which was expensive and affected the competitive factors of the mill's operation. In addition, most people, contractors and homeowners, did not care; they preferred trim that was not overly expensive, and very, very few could afford hand labor for such nonessential refinement after 1900.

As a consequence, most of the molding sold and used in modern buildings had become quite standard by the 1920s and '30s. Crown mold, today, offers the same ogee curve in Seattle as in Miami. Stop mold varies only in width for the most part. Brick mold, bed mold, and picture mold are the same in all parts of the nation.

The same relative lack of variety also applied to boards. There are a few standard widths and thicknesses, and the builder uses these because they are easily available and less expensive than boards with special dimensions would be. In Revolutionary times timber was often cut by pit saw and could be ordered in any dimension desired, though it must be said that standard dimensions were often ordered from the sawpit because they were satisfactory and easier to specify.

Perhaps the most revolutionary developments in house building in the post-World War II years have been the prefabricated house and its cousin, the premanufactured mobile home, or trailer home, delivered complete with fine furniture and suitable for either a sedentary or a nomadic life. In some ways the trailer home is as exemplary of modern mobile America as the log cabin was of early America; each was developed to fit the conditions of contemporary life. The main difference is that the log cabin reflected the hard work of the individual and was beautiful while the trailer home reflects an industrial America and is sometimes attractive but more often just clever. However, most modern Americans, given a choice, would choose the trailer instead of the cabin, regardless of the aesthetics involved.

A great laborsaver for builders since World War II has been plywood, upon which a large and profitable industry has been built. Using second-grade logs, rotary-cut into thin sheets which are laminated with modern adhesives, plywood provides strength, greater width than sawn planks, and easier installation for subflooring, sheathing, subroofing, and many other uses in modern building. Common plywood is not pretty, and is usually used where it does not show. Veneered plywood, however, is also available in a variety of woods and styles for quite handsome plain paneling.

Plywood would have been so expensive to manufacture by hand that its use would have been incomprehensible before 1900. It is a true product of the industrial age.

Many other products, which by the 1920s had been widely used for a couple of generations, have made the housewright's job easier. Asbestos shingles with fake wood grain are often substituted for the real thing; asphalt roofing has almost universally replaced wooden and slate shingles, thatching, and terra-cotta tile. Many modern houses utilize aluminum clapboards which seldom need painting. Wallboard, easily installed and requiring less skill, has replaced plaster keyed on wooden laths in even the most expensive houses.

Steel goes into more and more buildings, including residences. Many modern farm buildings and warehouses are built entirely of steel sheets fastened to steel frames for strength, providing minimum maintenance and safety against fire. Offices, which in a simpler America were once almost invariably a part of a house, are now in buildings made of steel and aluminum, never touched by claw hammer or plane or ancient adz.

Any house, no matter its type, size, or period, must be built with tools of some sort. But the tools that build America today are mostly impersonal, organizationally oriented tools. They can't be held in the hand or made by the workman. They consist of complicated machines in huge buildings and miles and miles of copper wire. These machines are manned by hundreds of people, each a specialist, who work in concert to produce a piece of molding, an "A" strut, or a window frame that will be incorporated into a house the workers will probably never see because it may be built thousands of miles from where they work. A factory is a tool, nowadays.

Certainly the housewrights still use hand tools, but they are greatly diminished in variety and most are power tools equipped with pneumatic or electric motors. Thus the craftsman, once dependent on an innate sense of rhythm and delicate skill, is now dependent on a power plant miles away and the built-in operational advantages decreed by the anonymous designers of his tools. The craftsman seldom designs his tools to fit his needs and his personality; his needs are researched by other people, his tools are supplied by others, and he must learn to use what is available. As a consequence his work reflects the organizational intellect of our times rather than the personal character of the worker. Our houses and furniture in the 1970s represent progress, but seem to have lost some of the civilizing influence of the individual craftsman and his art. But alas, shelter is a more basic need than art, and without modern tools such as factories and power saws, most of us might stand shivering in the snow as we admired the hand-built houses of a few fortunate, immensely wealthy citizens.

Man is still a weak, cowering creature when pitted against nature. His tools have been his shield and buckler and will continue to be so as they have been since proto-man first appeared in the game-rich savannahs of southern Africa some twenty million years ago.

But this book will limit its observations to the tools that built and will continue to build houses in America, a land that has contributed mightily to the concepts whence have sprung the impersonal tools of today. Its emphasis, however, will be on the old tools, the hand tools, the true reflections of individual man's ingenuity, indeed, his special genius for developing a system that creates and maintains the comforts of life.

The old tools have disappeared from our lives for all practical purposes. They are not likely to be revived unless every power plant vanishes overnight along with the host of engineers who design and build them and the considerable literature which teaches incipient engineers.

From our past aspirations, however, spring our future inspirations. The old tools should be remembered, studied, and admired as the simple but ingenious answer to the question of how to live as a human being. That question was particularly difficult in the simple beginnings of America, but tools gave the answer. They helped man use nature for his own comfort. Perhaps they subtly influenced Americans to become independent of King George III. Certainly they show that every man can live comfortably without dependence on anything but himself if he nurtures the right attitudes and creates the proper tools.

One-room log cabin

1

THE
LOG CABIN
AND THE PIONEER
JACK-OF-ALL-TRADES

If ever there has been a symbol of the early American frontier, it is the formerly ubiquitous log cabin.

It could be built quickly and relatively easily, entirely from the trees of American forests, and with a minimum of tools. When built well it lasted for generations, providing shelter against the most severe weather and serving as a sturdy fortress against Indian attacks.

Like most cultural expressions in America, the log cabin reflected European origins. It is thought to have been introduced to the New World by the Swedish colonists who first settled Delaware as the most far-flung outpost of King

Two-story log house

Gustavus Adolphus. Familiar also to the Germans who emigrated to Pennsylvania in the early eighteenth century, the log cabin soon was adopted by English and Scotch-Irish settlers as they populated the empty, forested reaches southward along the foothills of the Appalachians. These Britons, seeking in the New World a comfort and opportunity they had not known in the Old, adapted the construction of the Swedish cabin to the simple floor plans of the stone and timbered cots they had known for generations in England, Scotland, and Ireland. Often one could determine the Old World provenance of a man's ancestors by examining the plan of his cabin and the presence of dovetails or saddle notches.

Finer cabins with multiple rooms and stories might require a few more sophisticated tools, such as a mortising chisel and mallet for fitting floor and ceiling joists. An auger was necessary to bore holes for pegs and treenails which fastened door and window framing or provided a place to hang clothes. Since all log cabins in the old days had split shingle roofs, a froe and maul, the latter, invariably handmade, were necessities.

To minimize the number of trees needed for a cabin the builder usually split his logs in half with such simple tools as the iron wedge and sledge, or the wooden wedge, called a glut, which was driven in with a homemade maul. The German settlers of Pennsylvania introduced to the American frontier the holzaxt, a specialized splitting ax which was adopted by southern pioneers and given the name "go-devil."

So, equipped with tools that could be carried by one man, the American pioneer built his own houses almost entirely by his own efforts, and did such a good job that his cabin outlasted himself and even his grandsons.

AXES

Ancient in design, inefficient in function, the **felling ax** was the first weapon used by Europeans against the virgin forests of a newly discovered continent. It equipped the gentlemen woodcutters of Jamestown, the laboring pilgrims of Plymouth, and the mission Indians of St. Augustine. Relatively long-bladed and lacking a poll, this imperfectly balanced tool nevertheless enabled the colonists of England, Spain, France, Holland, and Sweden to carve a foothold and establish a new society. It was improved upon only in the middle of the eighteenth century by settlers who needed new tools to meet the problems of a vast frontier.

After a century or more of cutting huge trees and clearing land with a tool which had not changed since the Stone Age, American colonists anticipated the industrial age by modifying the ancient felling ax into a more efficient tool. Some backwoods blacksmith, somewhere, perhaps executing the concept of a weary woodcutter, produced an ax that had a shorter, broader blade and a poll attached to the eye opposite the blade. The result was an ax with well-nigh perfect balance that could be swung easily for hours with no effort required to keep the blade aligned with the kerf. Its poll provided more weight behind the edge, allowing the momentum to push the sharpened edge into the resistant tree; its wide, rather thick, blade made a bigger cut and more easily removed the chips, some as large as a dinner plate, from the kerf. With it an experienced axman could fell a two-foot tree in less than half an hour. The genius behind its design, mostly unnoticed in the sophisticated atmosphere of an industrial society, was reflected in its adoption in most countries of the civilized world. The building of America rested far more on this prosaic tool than on the more romantic symbols of long rifle and six-shooter and locomotive. Its rudimentary influence was probably more important to the future of America than any manned moonshot.

As America entered the industrial age around 1840, and the population exploded, creating a vast need for housing and transportation, mostly made of wood, the incipient lumber industry was strained to supply the country's needs. Maine, with its largely untouched forests, was where wholesale commercial timber cutting for the lumber mills began. There the

European felling ax

American felling ax

American double-bit ax

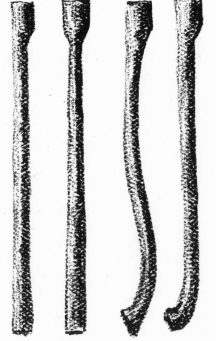

Shapes of ax handles

uniquely American polled ax was discarded for the **double-bit ax** that later was used almost exclusively to clear the forests of Michigan, Wisconsin, Minnesota, the Appalachian regions, the virgin pine forests of Georgia, and finally the heavily forested slopes of California and the Northwest Coast, where it has only recently been supplanted by the horrendous gasoline power saw. Actually the double-bit ax developed in Maine was a reinvention of a most ancient tool. Minoans and Romans used a double-bit ax in ancient times; and Danish woodcutters in the nineteenth century were also known to use a double-bit ax. Perhaps it was Scandinavian lumberjacks who developed the form in Maine. Regardless of its history or past forms, the American double-bit ax developed into a graceful, efficient tool, well-balanced and providing two sharp blades. With the crosscut saw, it was an important factor in the settlement of the Midwest and the northwestern sections of the country and in the building of America's cities.

Another peculiarly American modification of the ax was the curved handle developed for the single-bit felling ax shortly after the Civil War. These graceful appendages, known as colt's foot or fawn's foot handles, replaced the straight handles used on axes since prehistoric times. The curve, with a knot on the end, allowed improved control over the swing of the axhead and provided additional accuracy. The curved handle, like the polled head, has gained recognition through its acceptance all over the world. Straight handles, of course, continue to be required on double-bit axes.

Logs too large to be used for specific purposes in pioneer times were reduced by splitting to supply properly dimensioned timber. Fence rails, cabin logs, puncheons, corner posts for well houses, billets for shingles were all split by using a club to pound a wedge through the log. Iron wedges have been used since olden times, but where iron was scarce a wooden wedge, known as a "glut" in eastern America, was used. These wedges were forced in with mauls, clubs with a heavy head and fairly short handle chopped from a single log, or by beetles, a giant-sized mallet with a long handle, its head bound with iron hoops to prevent splitting under repeated

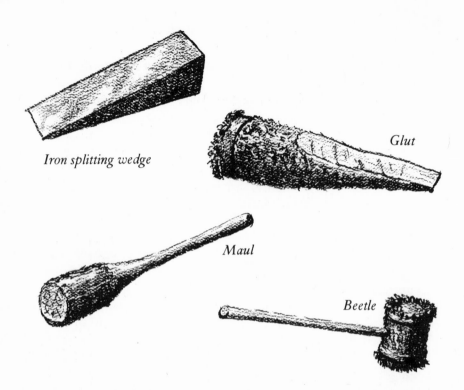

Iron splitting wedge

Glut

Maul

Beetle

pounding. Some splitters employed a steel sledgehammer for use with metal wedges, but the more careful preferred a striking tool which would not batter and spread the malleable iron.

In the Pennsylvania Dutch country a widely used splitting tool was the **holzaxt**, the name of which showed its pure German origin. At some time or another it drifted South and became known as a splitting ax or go-devil. With a six- to

Holzaxt or go-devil

eight-pound head, a heavy eye, and an obtuse blade, seldom sharpened, it could serve as an ax, a wedge, or a sledge. Its weight would split a good-sized log up to four feet long. For logs of fence rail length the go-devil might be used to start the split enough to insert gluts or wedges which could then be pounded in with the poll of the go-devil. Never a widely used tool, it is still made by a couple of ax manufacturers and is sometimes found for sale in modern hardware stores.

Split rail snake fence

Split fence palings

Split post and rail fence

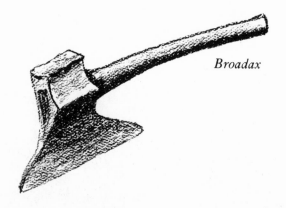

Broadax

There were several varieties of the **broadax**, all large and relatively heavy; all were employed in conjunction with the felling ax to square large timbers such as the logs with which a cabin was built. The name is obviously appropriate, for it designates an extremely wide blade suitable for planing a large area at one blow. Most broadaxes, but not all, had a heavy poll behind the eye which could be used to pound house timbers together at a mortise and tenon joint. Some, which apparently originated in Germany and were widely used by the Pennsylvania Dutch in America, were known as "goosewing" broadaxes because of the graceful ogee curve of the blade from eye to straight edge. Others, used exclusively for hewing, had a wrapped eye with no poll. Most, with the exception of the knife-edged broadax, used for rough hewing, were bezeled on the right side of the edge only, and had the eye protruding to the right to leave the left side perfectly flat for hewing with a long stroke. The bezeled-edged axes were usually equipped with a handle which curved to the right so that the axman would not mash his fingers between handle and log, which was located on his left. The broadax began to disappear it the 1840s, but became a manufactured item later in the nineteenth century when its use was revived in the South to shape crossties for the railroads. Broadaxes were used to flatten a log which invariably had previously been scored deeply at six-inch intervals with a felling ax. It is the marks of the convex edge of the felling ax which are seen on the hand-hewn logs of old cabins, or the sill timbers of old houses. A good broad-axman took pride in precisely hewing to the line, which he marked on the timber with chalk or charcoal dust applied with a snap line, making his broadax serve as a gigantic swing plane.

Logs squared with a broadax

Brake for riving boards

SPLITTING TOOLS

In the beginning virtually all the roofs of all the cabins in America were made of shingles split from oak or pine, chestnut or poplar, durable cedar and cypress. Clapboards were split, also. Three simple and ancient tools were used for splitting; the iron **froe** with a wooden handle, the wooden **maul**, or mallet, and the **brake** made from a tree fork or improvised from a couple of small straight logs. Even such a simple tool as the froe had to be used with considerable skill and careful technique. Each bolt of wood from which the board was split required a certain amount of study of its individual character. A good boardmaker, however, could turn out a couple of thousand shingles, or that longer type of shingle called a shake, or several hundred six-foot clapboards a day. Probably the first real organized industry in the first colony of the British Empire, Virginia, was the splitting of clapboards to be exported to England. The timber resources of the Virginia tidewater provided a welcome substitute for the oak forests of early seventeenth-century England, which were being systematically raped to make charcoal for the iron industry and to

Riving maul

Froe

A split shingle

supply timber for the English navy. With no sawmills established at Jamestown, splitting was the only suitable technique for making small boards. Even after sawmills were commonplace, most householders in the backwoods considered split shingles far more satisfactory and durable than sawn shingles.

Split shingles stacked

A roof of split shingles

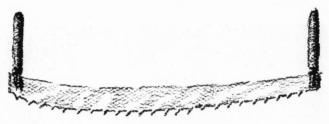

Two-man crosscut saw

OTHER TOOLS

Though not essential, a good crosscut saw enabled the pioneer jack-of-all-trades to cut the logs for his cabin more easily than by ax alone, and it made the jobs of trimming the ends of logs for neat corners, and shaping dovetails, much quicker and easier. The crosscut saw, its tooth pattern modified in the nineteenth century to make it more efficient, was also widely used by the logging industry for cutting the timber of Maine and the Northwest. Most crosscut saws were designed to be used by two men, but one-man saws became available in the late nineteenth century.

The **adz**, as old as the ax in concept, is sometimes described as a chisel with an ax handle, and at other times designated an ax with a perpendicular blade. Both descriptions may apply, for the function of an adz is that of a chisel, while the technique of its use is similar to that of the ax. Most backwoods households had a foot adz as part of the basic collection of tools needed to provide minor comforts and refinements to

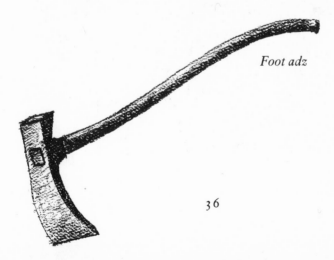

Foot adz

End-sawn cabin logs and sawn notches

An adzed board

An adzed ceiling joist

a home isolated in the wilderness. It was kept razor sharp and was used to smooth floor puncheons, or to make table tops, or to provide any type of board which needed a smooth, level surface. Adzing was a dangerous occupation for any but the most skillful. The tool was used by standing on top of the log to be smoothed and swinging the adz precisely to create an even plane of the wood below the adzman's foot. If it was not used precisely the razor-sharp blade would easily cut through the shoe to the flesh and bone of the foot. Nevertheless, some adzmen were so confident of their skill that they would win bets that they could split their shoe sole without damage to the foot which was placed a fraction of an inch above the callous steel.

In the days when nails were scarce and pegs were used as a substitute, the **auger** was absolutely essential for boring the holes into which the pegs were driven. Most pioneer home-steaders had augers of several sizes in their tool chests. Before 1800 augers were of the spoon, or pod, design, which required considerable pressure to force the curved, sharpened end into a log or board. After 1800 the modern twist auger became more and more common. The twist auger was much easier to use by virtue of its screw point which pulled the bit into the wood, and more precise because the twisted bit removed the wood shavings and controlled the direction of the hole. Augers were not only used in building cabins, but also for starting the mortises in post and rail fences and in making sleds and other pieces of farm equipment made of wood.

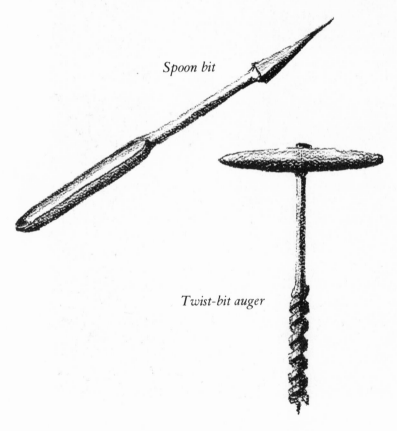

Spoon bit

Twist-bit auger

Auger-bored mortise in a fence post

Legs fitted in round mortises bored by an auger

A mortised timber frame in a brick wall

Although many a log cabin was built after 1840, that year is generally recognized, somewhat arbitrarily, as the beginning of America's industrial age, give or take ten years. It was around then that hand tools slowly began to be discarded and replaced by the tools of impersonal mass production; sawn lumber, for so many centuries the luxury product of the hand sawyer, became a commonplace building material; and mass-produced cut iron nails became a commodity, replacing the expensive handwrought nails, which had been used since biblical times.

It is true that mill sawn lumber, the product of inefficient, up-and-down sash mills run by water, had been familiar for several hundred years at the time America was first colonized. About 1850, however, the concept of the circular saw was realized. Steam power freed the sawmills from thralldom to the streams, for the new steam engines of the times could be transported, however tortuously, to any place where timber was found. After 1850 one found sawmills even in the flat country of the coastal plains and on the seacoast itself where slow meandering streams were unsuitable as a source of power for sawmills. Later still, the internal combustion engine provided even more transportable power so that a mill could be used one day in one area and easily disassembled and set up the next day at a location miles away from the first.

This meant, of course, that a farmer wanting a new house could fell only half the trees needed for a cabin and these trees would yield enough rough sawn boards to make a house of several rooms and a floored porch. After World War I, a backwoods farmer who had been reared in a log cabin could even convert his Model-T Ford into a sawmill and produce his own lumber. So the circular saw joined the devices that eventually separated the American freeholder from his long, personal relationship with hand tools.

The change was not apparent day by day or year by year, but it was plainly apparent generation by generation. The first generation away from the log cabin often built a clapboard house of frame construction, ceiled on the inside with beaded boards, roofed with sawn shingles or sheet iron. Such a house was generally a symbol of prosperity and, since it had a number of rooms, a symbol of status.

Those less prosperous might sheath their smaller cabins with board and batten, ceil the inside and put on a split shingle roof. Later still, when roll tarpaper became prevalent, the sheathing might be covered with this somber material, the battens then being applied to fasten the paper down. Often the roof, too, would have tarpaper substituted for shingles. Many houses of board and batten construction are still to be found, usually abandoned, in farming areas around the country. That most hideous of synthetic materials, asphalt sheathing with a brick pattern, was used as sheathing on some.

Of course, the tools used in building latter-day equivalents of the log cabin changed greatly in character, and so quickly that many of the old tools of true frontier days virtually disappeared. The broadax and adz were no longer needed when working with boards which came from a nearby portable sawmill, nor was a twibil, nor auger, nor maul and froe. About all that was needed, in fact, was an adz-eye hammer, which appeared about 1840, handsaws for crosscutting and ripping, and kegs of nails—cut nails until the 1870s and after that the cheaper wire nails developed in Europe and soon adopted wholesale in America. Also, the more modern housebuilder often needed a plane or two to trim door and window facings to fit, and of course he used levels and squares which could be easily carried in a wooden box, homemade, which he slung over his shoulder.

One finds a cabin class of American society even during the period of affluence following World War II, but the cabins of the space age are for the most part shell houses, prefabricated houses, and mobile homes. They are many steps removed from the personal efforts of the owner, and built of materials harvested or manufactured long distances from the site. Many of them are well built but they lack the spiritual quality of a family-built house. Most are built of wood, but the heavy work is done in faraway factories, the parts cut and assembled according to preconceived plans instead of being designed to fit circumstances of individual tree trunks and limited tools, factors that made the old log cabin a work of art.

In the factories, new and complicated tools are employed. Huge bench saws cut already planed boards. Even the ancient hammer and nail is often discarded for pneumatic or electric staple guns. In some instances new adhesives obviate the need for nails.

On the site, carpenters work only as assemblers with hammers and power saws. Precut walls are nailed together, already assembled door and window units are nailed in place, preassembled struts form the roof. Sheathing and subflooring,

usually of great sheets of plywood, are nailed on and clapboards already cut to length are nailed to this. So simple is
the assembly operation that the carpenters no longer need
even a tool chest. They raise houses in one day with hammer
and saw, level and square. A modern carpenter might work
on several hundred similar houses in a year's time. He can
only reach such volume of production with new tools suited to
his needs.

It is estimated that the electrically powered circular saw
allows one man to do the work of ten men using hand saws.
Such modern equipment is fast, but it's also dangerous. The
modern carpenter-sawyer must keep his wits about him, and
pay as much attention to safety measures as his eighteenth-
century counterpart did to sawing a straight line. But how
else could houses be built in sufficient quantity in the 1970s?

Ah, Ben and James and Eli, we pay homage to your genius,
but what have you done to our lovely log cabins? You've
made them alike; all alike, faceless and square.

THE
HOUSE
AND THE
CARPENTER

Most houses of pre-Revolutionary America were built of wood, and a large number of modern houses in America are built of the same material. Even structures built of brick and stone in the old days, or of metal and plastic and patent compositions in modern days, have a great many wooden components, such as floor and ceiling joists, doors, windows and their frames, sheathing, and subflooring.

A multistoried house built entirely of stone, for instance, requires heavy vaulted ceilings and roofs, constructed with great labor and needing complicated wooden forms for support. Steel, aluminum, and cast concrete have replaced wood in some modern structures, but these materials are rather expensive and difficult to use for residences and are found mainly in large commercial buildings.

So, wood always has been and possibly will continue to be an important structural material in residences. Certainly it was important to the British-born ancestors of the colonists who first settled Virginia and Massachusetts, and it was needed to some extent even in the brick and adobe structures of the Spanish settlers of Florida and the Southwest. Wood was the most available building material in most regions of America, far more easily accumulated and processed than quarried stone or burnt brick.

The earliest American houses, the types found in all sizes, in towns and on prosperous farms, were considerably more complex in construction than either the log cabin or most modern houses. Each component was shaped carefully and the separate parts were fitted together almost as precisely as the parts of a piece of fine furniture.

The old-time master carpenters set up a regular workshop in an unfinished house, brought in a remarkable variety of saws and planes and adzes, and formed every board by hand, thin shaving by thin shaving; saw cut by saw cut. Houses had to be built to last, for the labor involved in building just one was enormous.

An outstanding example of the durability of old building techniques is an ancient tithe barn built in England in the eighth century, and still standing. The basic techniques used in building this barn were still being used by American Colonial housewrights some eight hundred years later, and the techniques were not discarded until around 1840. Some American houses built in the first days of the Colonies are still standing and some of them are still inhabited. Those that have not survived disappeared mostly because of fire or cannon or acts of God, not because the builder was at all derelict.

There was little change in the basic technique of building houses, large or small, from about 1840 until 1940. After World War II, however, industrialization hit its stride, and an astounding proportion of very good single-family residences built since 1940 were planned and built piece by piece in factories and assembled on the site after the foundation, often a mere concrete slab, had been prepared.

A modicum of handwork is still needed for assembling a precut or prefabricated house. All carpenters still use that ancient tool, the hammer, and most use a plane or two and a handsaw or two. But most also own and generally use a power saw and many will set up an electric bench saw, with planing attachments, at the house site, following perhaps the spirit of the eighteenth-century master carpenter if not his methods.

There was some industrialization in the lumber business before the American industrial age was born. Near the larger towns one could usually find a water-powered sash saw, or up-and-down sawmill, which provided sawn boards for town houses with a minimum of manpower and with many times the production capability of the pit sawyers, those men who patiently sawed boards from huge trees with large two-man ripsaws. The sash saw was a welcome labor and cost saver,

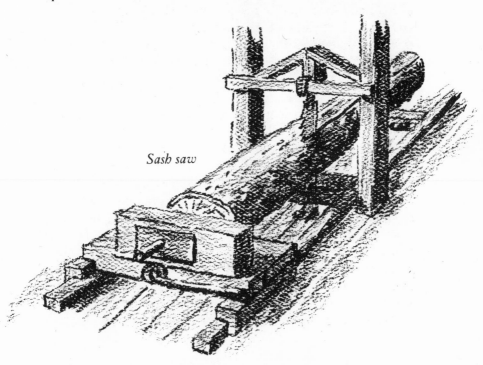

Sash saw

A sash-sawn timber

and used wherever available. Log cabins of post-frontier days, for instance, were often assembled of hand-hewn logs and floor joists, but with sash-sawn ceiling joists and trim. (In the South, however, many of the fine plantation houses had frames of hand-hewn timbers until well into the industrial age, but for economic rather than aesthetic reasons. For the cotton-farming entrepreneurs of the South had access, at their house sites, to all the timber they needed as well as unlimited slave labor. It was cheaper for the Southern plantation owner to use hand labor and hewn timbers, despite the waste of hewing.)

Circular saw

When houses began to be built wholly of lumber sawn and finished at the mill the skills and tools of the carpenter declined. His ancient tools, however, when recognized, still serve as symbols of industry and ingenuity. Each of his tools was a source of comfort and protection in a preindustrial society.

A board sawn with a circular saw

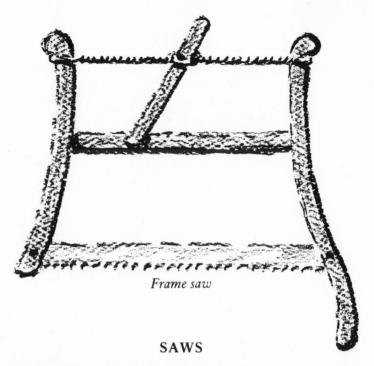

Frame saw

SAWS

Handsaws have been part of the carpenter's tool kit since the Bronze Age and before.

There are two basic types of handsaws. **Crosscut saws,** designed to cut across the grain of a board, are merely a series of small pointed knives which slice through the alternate hard and soft fibers of the grain. **Ripsaws,** on the other hand, are a series of small chisels designed to scoop out the wood in the direction of the fibers, literally ripping a cut down the length of a board.

Frame saws were a most logical form in the days when steel was scarce and expensive and its quality varied from inch to inch in every saw blade; the frame kept the thin blade taut. Because of the difference between cutting crossgrain or ripping, frames were made in two different forms.

Crosscut saws had the blade mounted in the ends of two strong boards. A wooden stretcher was placed midway on these boards and the ends opposite the blade were drawn together with a loop of cord. The cord was twisted by means of a short stick, thus drawing the two loose ends of the frame together and consequently separating the opposite ends to stretch the blade. When sufficiently tight, the short stick inserted in the cord was stopped by having one of its ends rest against the stretcher. Such a frame was quite satisfactory for sawing a board in two across the grain. This type of frame made of steel, survives in modern times in the coping saw and the metal cutting hack saw, both of which have thin, narrow blades.

For ripping boards over two feet in length, however, the crosscut-type frame would soon interfere with the stroke. Ripsaw blades, as a consequence, were mounted in a rectangular frame, the blade being fastened at each end in the center of handle boards from one to two feet long, with two parallel stretchers placed between the ends of the handle boards. The

A board crosscut with a frame saw

stretchers were placed far enough from the blade to clear the edges of a board being ripped. The blade was tightened with thumbscrews or small wood or metal wedges.

All frame saws generally fell into these two categories, from prehistoric times until the middle of the nineteenth century when frame saws became obsolete except in one or two forms. During the seventeenth century, the Dutch, experts in steelmaking, designed a wide-bladed handsaw with a crooked handle, somewhat resembling an umbrella handle. This type, called a panel saw, the forerunner of the handsaws still available in modern hardware stores, was further developed in England during the eighteenth century. The chief improvement on it was an open handle which allowed greater control of the angle of stroke, and more comfort for long hours of sawing.

A large, heavy two-man **pit saw** was needed to rip a log into boards, with one man standing below, and the other standing on top of the log being sawn. Sometimes the log was positioned over a deep pit, hence the name of the saw and the technique. Usually, however, the log was placed on a scaffold, with the pitman on the ground and the sawyer on the log.

In early times, pit saws were invariably mounted in a frame. Later, toward the end of the eighteenth century, when high-quality cast steel became available, saws were made without a frame with a T handle riveted to one end of the long blade and a slotted block of wood fitted over the blade to serve as an easily removed handle on the lower end.

Usually the pit saw was designed to cut only on the downward stroke, the weight of frame and saw doing much of the work. A few were designed to cut in both directions.

One other type saw, with origins in Roman times at least, was also used by the Colonial American carpenter. This was the **backsaw**, a handled saw with a rectangular blade which was made stiff by a strip of iron or brass folded over its back. These were used to cut precise miters, dovetails, mortise shoulders and butt joints where the flexibility of a saw blade might ruin the straightness of a cut on a fine stairway or mantel.

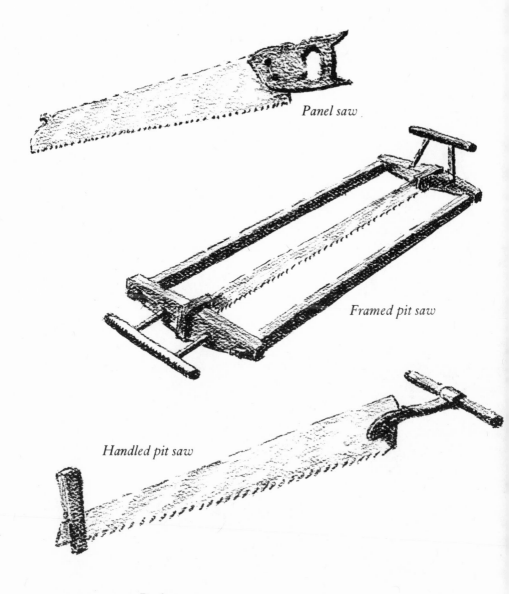

Panel saw

Framed pit saw

Handled pit saw

Backsaw

A pit-sawn board, showing irregular saw marks

A board ripped with a panel saw

A mitered corner cut with a backsaw

HAMMERS

A Roman carpenter would easily have recognized any hammer used by an American carpenter from the earliest Colonial times up to the present time. This basic carpenter's tool is known as a **claw hammer** because of its curved, split poll which is used for extracting nails. For centuries the claw hammer was no more than a short bar, face on one end, claw on the other, a hole punched in its middle to receive a wooden handle, almost always of hickory in America. In the days of mortised construction when nails were sparingly used this simple tool sufficed for hammering and nail pulling. In nailing clapboards, however, the handle, having no depth beyond thickness of the bar, usually became loose, or broke, when used to pull bent nails in any quantity during a day's work.

Often the carpenter would have the blacksmith weld iron straps two or three inches long on either side of the eye, to extend along the handle and be fastened to the wood with screws or rivets. This, of course, made a better tool for nail-pulling, but the straps inhibited the natural springiness of the hickory handle and made the tool poorer for driving nails.

This problem was intensified about 1840 when balloon construction of houses and consequent greater use of nails was adopted. David Maydole, a blacksmith of Norwich, New York, at that time invented the adz-eye claw hammer, adopting from the ancient adz an elongated eye which stiffened and strengthened the wooden handle but did not detract from its springiness. In a few years the adz-eye claw hammer was adopted eagerly by carpenters all over the world. It has not been improved on since its invention, although many modern carpenters prefer a hammer with a tempered steel handle integrally forged with the head, making it impossible to separate the two components. This type was developed in Rockford, Illinois, in 1926.

Of course, as houses have become an industrial product, made in factories, the ancient hammer has often been replaced by pneumatic nailing machines, which not only drive nails but actually make the nails from long coils of wire with

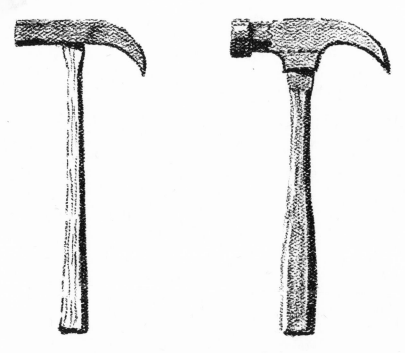

Hammers—both old-style and adz-eye

which the machine is loaded periodically. Plasterboard and thin trim in modern houses are frequently installed with a spring-operated stapler, much faster and more productive than a hammer.

NAILS

The only *raison d'être* for claw hammers, of course, has been to extract the metal nails as fastening devices in wooden structures. There have been nails quite as long as there have been claw hammers, in a variety which defies accurate cataloging. Nails, or pins, are referred to in the Bible in the description of Solomon's Temple. Numbers of Roman nails have been excavated in camp and town sites all over Europe. And while they were scarce and expensive during the Dark Ages and medieval times, the nailer, who made nails, was a recognized tradesman for centuries before America was colonized. But nails were used sparingly in the New World until after the American Revolution.

For the most part nails were made in the regular blacksmith shops, usually by an apprentice who was assigned the boring task as part of his professional training. Sophisticated shops near good transportation bought nail rods of various sizes from the iron furnaces and rolling mills of the day, and made nails with forge and hammer and header, sometimes as many as two thousand a day per nailer. Country shops, isolated from transportation, often set apprentices to drawing out worn horseshoes into quarter-inch square nail rods, then to transforming these into the type of nails needed in the community. Throughout Colonial history, however, and until the end of the eighteenth century, America imported quantities of nails from Europe where they could be made at low cost by professional nailers who used special nailing anvils and other exclusive equipment. Toward the end of the era of handwrought nails, Thomas Jefferson operated a nail manufactory as part of the plantation-industrial complex at Monticello.

Handwrought nails came in many forms. The most common was the **rose-headed nail,** used for fastening clapboards and shingles and for installing hinges on doors before the days of wood screws. The **finishing nail** was made with an inconspicuous T or L head which could be driven into floorboards and wallboards so that the head was practically invisible. **Brads,** the eighteenth-century term for tacks, had large, flat, broad heads and were used to fasten cloth or leather to wood. Small finishing nails were used to fasten molding to panelling.

Most handwrought nails had a sharp point forged on a square shank. As a consequence they wedged between the fibers of the wood and were practically impossible to remove.

Around 1790 the **cut nail,** as revolutionary in concept as American theories of government, appeared. Cut nails were invented by several individuals simultaneously in the years

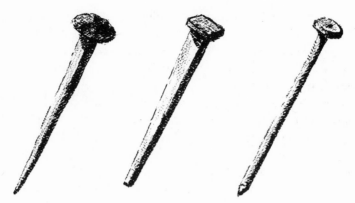

Three types of nails—wrought, cut, and wire

immediately following the Revolution, and in 1790 several people started factories to compete with the handwrought nail factories that had appeared in New York and New England after the war when imported European nails were difficult to acquire. Within a few years the cut nail, mass produced and inexpensive, had almost totally replaced the handwrought nail in America. By 1800 new machines had been designed to cut several varieties of nails of many sizes, but the factories, as usual, extinguished the vast differences found among handmade nails. Strangely, carpet tacks continued to be made by hand for another century after the cut nail appeared.

Originally cut nails had the fiber of the wrought iron across the point of the nail, a feature that weakened the nail. Heads of cut nails originally were formed by hand, but since this operation was done cold the heads were mere vestiges of the beautiful rose heads which preceded them. Later machines were developed that headed the cut nails automatically. By the 1820s cut nails were made with the iron fibers running the length of the point.

Wire nails, the type most common in the major proportion of buildings now standing in America, were developed in England about 1850. They were not widely used in America

Hand-wrought rose-headed nails in an 18th-century door

until the 1880s and 1890s, but once American industry began to produce them the wire nails more or less completely replaced cut nails because of their cheaper manufacturing costs. Cut nails, however, are still readily available today in a few sizes.

Since World War II more and more residences are being fastened together with wire staples and various types of synthetic adhesives. Most of such fastening is hidden in various components of prefabricated houses and not readily apparent to the naked eye. Today nails are still common in all residential buildings and they are not likely to disappear very soon. Undoubtedly, though, new types of fasteners will be used in greater proportion in future years, and the ancient claw hammer will be used less and less.

Hand-wrought T*-headed nails in an 18th-century floor*

Cut nails in the door of an 1830 house

HATCHETS

Before the industrial age, when a housebuilder was called upon to do much of the work now supplied by mills, hatchets in several forms were quite necessary items in every carpenter's tool chest.

The regular hatchet, of about two pounds weight, was no more than a miniature broadax with a short handle. Its edge was bezeled and the eye offset on the side of the bezel with the opposite side a plane, to allow hewing a straight line. The careful workman kept his hatchet as sharp as an adz, for it would not cut dependably otherwise. Hatchets were used for roughhewing generally, on boards which would not be seen or which would be planed after hewing. Some modern toolmakers still offer hatchets, but few have been bought since World War II, and then usually for trimming warped joists or wall studs which interfere with placing flooring or wallboard.

Similar to the hatchet, a **half-hatchet** blade was flared only on the bottom and was straight across the top. Frequently the poll was drawn out into a hammer bit and both the old handmade specimens and later factory-made types generally had a notch on the bottom of the blade for nail pulling. Half-hatchets were used for the same purposes as the hatchet, but were also handy for nailing, splitting, trimming, and tearing off old wooden shingles before reroofing a building.

Carpenters often employed a more specialized tool for shingling known as the **shingling hatchet**. It was a light version of the hatchet, but with a knife edge instead of a bezeled edge.

Another specialized form was the **lathing hatchet**, with a narrow knife-edged blade, hammer poll, and nail-pulling notch. This type was flat on top so that the hammer could be used to nail laths in the corner of wall and ceiling. The narrow blade was used to trim and split laths to fit in certain places.

Possibly the lathing hatchet, which is still available in a few hardware stores, was the last type of hatchet to be developed by carpenters. It has almost disappeared, however, as

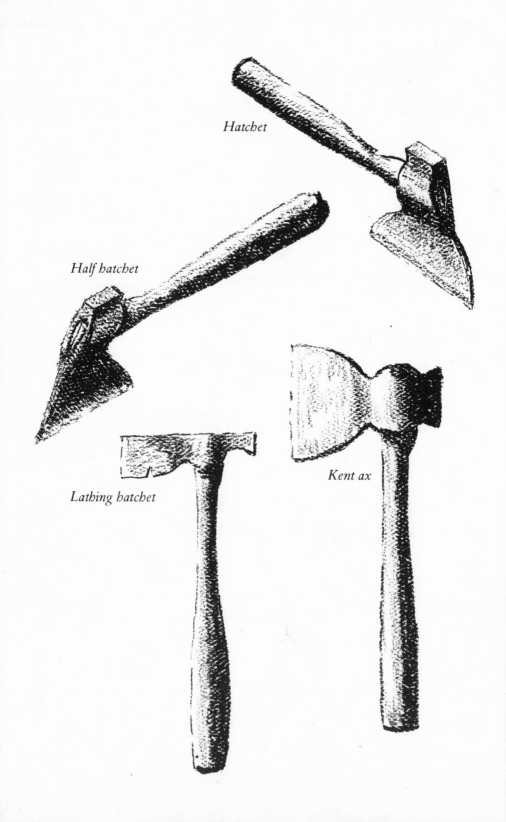

Hatchet

Half hatchet

Lathing hatchet

Kent ax

A board roughhewn with a hatchet

has the half hatchet and shingling hatchet. Composition roofs, wallboards and mill-cut lumber have made the hatchet obsolescent.

Another tool, now obsolete and often not identifiable, was the **Kent ax,** a small hand ax with a special shape to its head, which was developed in England during the eighteenth century and adopted widely by American carpenters. The Kent ax had a straight edge sharpened like a knife and it had a light poll behind the eye. It was used for roughing and for pounding and was retained by Pennsylvania rural carpenters until the early years of the twentieth century.

Split laths in an 18th-century house

PLANES

When George Washington ordered planes from England to equip his plantation shops before the Revolution, he ordered them in sets of fifty, encompassing several types and numerous sizes. Of all the woodworking tools used from Colonial times in America until World War I, the family of planes was by far the largest category. An eighteenth-century carpenter doing common work was required to have from ten to twenty-five different planes. Master housebuilders and cabinetmakers required many more.

The basic function of a plane is to cut a smooth surface, a geometrical plane, on rough boards as they come from the sawpit, the sawmill, or the broadaxman. To accomplish this the tool itself was essentially a block of wood with a smooth bottom, the block pierced to receive a long blade sharpened on one end and held tightly in the aperture with a wooden wedge. In effect the block was a movable jig, its smooth bottom guiding the razor-sharp edge to shave off any irregularities in the surface of the board being worked.

Before the days of planing mills, carpenters used their planes to smooth all the boards used in a house. As a consequence, even the most poorly equipped carpenter had to have at the very least three different types of planes, and possibly additional sizes in one of the types.

Jack planes were from twelve to eighteen inches long and after 1700 usually had a closed handle similar to the saw handle developed in England in the eighteenth century. Since they were used for rough, preliminary planing, the jack plane generally had a slightly convex edge to its blade which enabled it to cut more deeply. The result was a wavy, rather than a completely flat, surface. Boards prepared with jack planes are often found in drawer bottoms in antique furniture and on the invisible back sides of old panelling.

Boards that had been rough-planed with the jack plane were then smoothed with a **trying**, or "truing," **plane** which had a stock from two feet to three feet long and a straight-edged blade. The longer stock provided a longer guide surface for long boards.

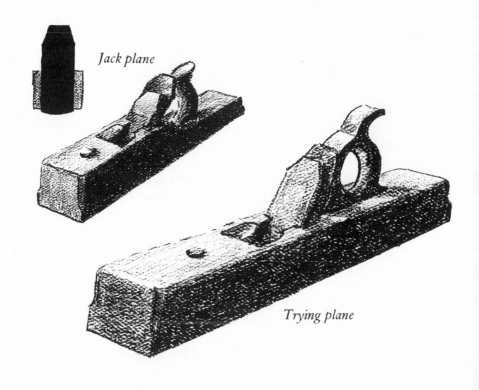

Jack plane

Trying plane

A board smoothed roughly with a jack plane

Door boards that were smoothed with a trying plane

Another similar tool was the **joiner's plane** with a stock thirty-six inches or more long. This was used mainly by the specialists who prepared intricate panelling which had to be joined perfectly. Some house carpenters used it also, under the name "floor plane," to level the surfaces of floors already laid with wooden pegs.

A floor leveled with a joiner's plane

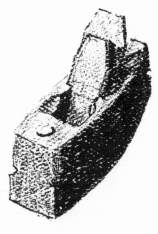

Typical smoothing plane

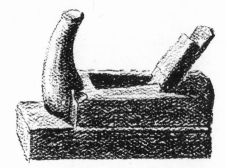

Horned smoothing plane,
common to northern Europe

The small **smoothing plane**, usually no more than eight inches long with a boat-shaped rather than rectangular stock, was probably used more frequently by the common carpenter than any other. Indeed, before the seventeenth century it may have been the only plane found in a carpenter's chest. Older smoothing planes frequently were equipped with a "horn," or upright curved handle, inserted or carved in the stock before the blade. Most, however, had no handle at all. This small plane was used to smooth the edges of boards or the sides of short boards. One of its principal uses was to chamfer panels for doors, walls, and shutters. When used for this purpose, though, the regular smoothing plane was used only to cut along the grain. A plane with a skew blade, set at an angle across the stock, was much more satisfactory for chamfering across the grain.

*A cabinet door with hip panel
formed with a smoothing plane*

Other types of planes, almost as archaic in origin as the smoothing planes, had different functions which belied the names of the tools, for they were designed to cut grooves, not planes. They are called **rabbeting planes** from the Old French *rabattre*, an ancient word meaning to beat down or reduce.

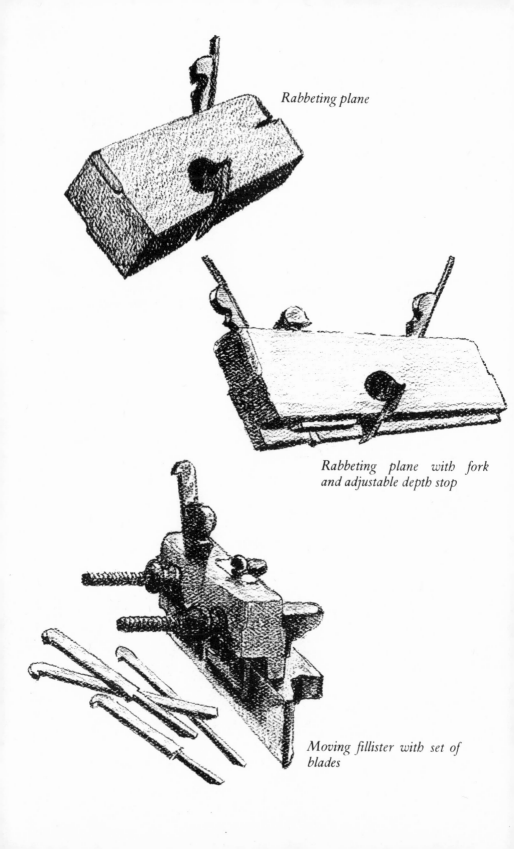

Rabbeting plane

Rabbeting plane with fork and adjustable depth stop

Moving fillister with set of blades

The simplest type had a blade which extended through the sides of the stock, so that the sharpened corners of the blade could cut corners in the stuff being worked. It was used to cut offset surfaces, or rabbets, such as the recesses of window muntins into which the window lights (panes) fit. This common type frequently had a thin board called a fence, screwed or tacked to one side, extending past the bottom to serve as a guide along the edge of the stuff. Another thin board could be fastened to the side of the stock opposite the fence, its lower edge a fraction of an inch above the bottom of the stock, which was called a stop. The stop controlled the depth of the rabbet being cut.

During the eighteenth and early nineteenth centuries plane-makers designed rabbeting planes with adjustable stops operated by a thumbscrew. A special type, mainly used for cutting crossgrain, had two blades, the front one a fork with its sharpened edges parallel to the sides of the stock, for outlining the groove to be cut, the other a skew blade with its edge in the normal position to cut between the lines made by the fork.

Rabbeting planes with narrow stocks and blades were also used to cut the channels in the middle of door rails and posts into which hip panels fitted. In the seventeenth century, when the golden age of woodworking began, craftsmen in wood developed the marvelous moving **fillister**, a rabbeting plane with an adjustable fence and, later, with an adjustable stop, which accommodated a set of blades of varying widths.

Ship lap joint formed with a rabbeting plane

A picture frame rabbet cut with a moving fillister

The most specialized of the rabbeting planes was the matched set of tongue-and-groove planes used to prepare the edges of floorboards, wallboards, and ceiling to effect a tight, windproof joint. Tongue and groove joints apparently were not used much until the latter part of the eighteenth century, but they became standard in houses from 1800 until the modern age of plywood and plastic tile floors.

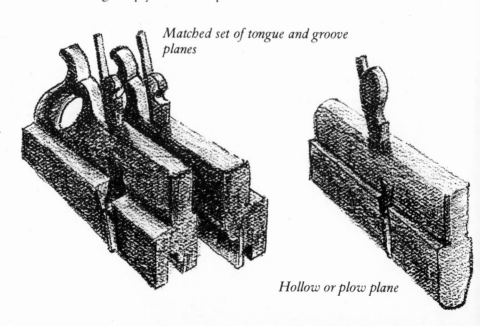

Matched set of tongue and groove planes

Hollow or plow plane

Possibly the most interesting of the planes used by the carpenter were the varied and versatile **molding planes**. Since about 1850 the molding found framing doors and windows and around the corners of walls and ceilings, as well as on the edges of mantels and bookshelves, has been commonly referred to as millwork, since it was produced by a planing mill. Before the industrial age, however, all such decorative trim was shaped by hand, using molding planes on the house site. The master carpenter usually did such work, with assistance from one or two skilled journeymen carpenters. Each of these artists might own a set of as many as a hundred planes.

The most common types were called **plows and rounds**, usually made and sold in matching pairs. Plows carved a concave groove in a board; rounds, or forks, a convex surface. Various sizes of plows and rounds used in artistic combination could produce very intricate molding of infinitely varied designs.

Beading along the edges of wall and ceiling boards to hide the joint was done, naturally enough, with the **beading plane** throughout the nineteenth century. As with rabbeting planes, the beading plane was often supplemented with a removable fence.

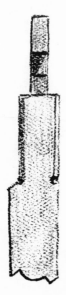
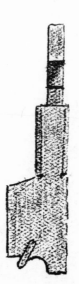
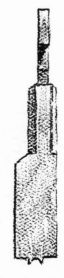

*Round or
forkstaff plane*   *Cove plane*   *Door mold plane*   *Beading plane*

Tongued and grooved board

Cove molding on an old bookcase

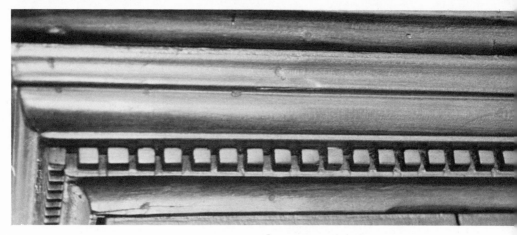

Eighteenth-century cabinet molding made with hollow and round planes

Beaded boards on an 18th-century batten door

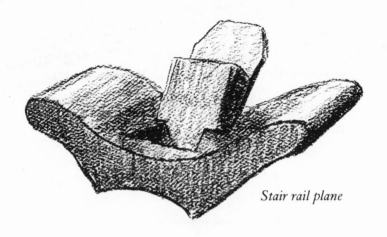

Stair rail plane

Crown mold plane

Crown mold and paneling on an 18th-century fireplace

Hand-planed stair rail

There were several additional specialized planes used by master carpenters for work which bordered on cabinetmaking, and, indeed, these planes were also used by the cabinetmaker.

One of these was the **chamfer plane**, a simple device designed to create a neat chamfer on posts as well as other objects. It consisted of a block with a V notch running its length, and a throat in which a blade, of the type used in a smoothing plane, was wedged. When in use the sides of the notch served as stops. The depth at which the blade was wedged determined the depth of the chamfer.

An auxiliary piece of equipment used in planing straight edges on boards was the **shooting board**, invariably homemade. It was made of two straight-edged boards, one perhaps five or six inches narrower than the other. The narrow board was glued, screwed, or nailed on top of the other, the back edges being together, the front edges being offset. A board to be planed was secured by holdfasts or clamps to the top board, its edge protruding slightly over the edge of the top board. To plane the edge the carpenter would place his short joiner on its side, blade toward the board being planed, and push it along the bottom board until its cutting action was stopped by the edge of the top part of the shooting board.

Chamfer plane

Shooting board

Chamfered street lamppost in an 18th-century town

During the seventeenth and early eighteenth centuries, when molding and panelling and the technology for tight joining was redeveloped, many carpenters made their own plane stocks of the simpler kinds. Of course, they ordered the irons, the blades, from the local blacksmith upon whom the woodworkers were dependent for the iron and steel members of all their tools. Planemaking became an established trade, however, in the eighteenth century, and most American tradesmen ordered sets of planes from Europe until the American Industrial Revolution began about 1840. After that most of the cities of the northern United States boasted communities of planemakers who furnished this important finishing tool, this source of woodworking refinement, to the carpenters and craftsmen building houses on the new farms and fast-growing towns of the West.

Wood-stocked planes prevailed in the carpenter's trade well into the 1890s with some improvements. For instance, in the late eighteenth century plane blades began to be made in two parts, the regular old-fashioned blade supplemented by a chip breaker which screwed onto the blade and broke the shaving as it emerged from the throat of the stock. This new improvement was called a cap blade and the planes were referred to as **double-ironed**. About the time of the Civil War, however, both the English and Americans developed wooden planes fitted with iron soles, an unconscious reversion to the ancient iron-soled planes of Roman times. Of course, by 1860, metalworking technology had progressed to the point where solid iron planes could be mass-produced at low cost. A few years later those tool manufacturing companies which had been established for thirty or forty years offered smoothing planes with full iron stocks and certain mechanical improvements. Also developed were ingenious stocks which could be adjusted to use twenty or more irons to cut molding, rabbets, tongue and groove, all with the one metal stock. These marvels evolved from the old wooden-stocked moving fillister. Most master carpenters of the 1880s and '90s owned one and maintained it as a prize possession.

By World War I all had changed. Most large modern

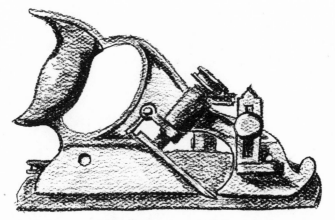

Combination plane of the 1890s with iron stock

building supply houses and lumber mills had by this time installed planing mills, and many of the larger mills provided moldings as well as finished boards. Part of the lumber industry specialized in oak flooring, delivering it all tongued and grooved and ready to install with hammer and nails, requiring only the limited use of a handsaw to trim the length to fit the room.

After World War II only finer houses used any molding at all, and the thousands of apartments built between the 1950s and the 1970s usually dispensed with crown mold, base mold, even door and window mold. Oak flooring has been largely replaced with plastic tile and wall-to-wall carpeting. The carpenters who work on these modern residences need only power saws and hammers; few of them know, or need to know, how to use or maintain even the most basic form of plane.

ADZES

The older a house is in America the more likely it is that its joists and some of its other timbers were shaped by a foot adz,

Carpenter's hand adz

just as with the log cabin. House carpenters before 1830 or 1840, however, almost invariably included a hand adz in their tool chests. Hand adzes were useful for all manner of odd jobs in putting a house together. They could be used to level floor joists before the flooring was installed, or to smooth studs before panelling was put up, and for other hewing that was difficult to accomplish with a hatchet. The hand adz was particularly useful in shaping curved corner braces or the mantels of window frames that fit into curved brick window arches.

But while the hatchet is still found in carpenter's chests today, hand adzes have been obsolete for so long that the few which turn up in America are usually misidentified as a form of mattock used for digging gardens. After all, joists as they come from the mill no longer require leveling, and heavy boards can more easily be shaped to a curve with an electric-powered saber saw. There is no place for the archaic hand adz in building a modern house. Today it is used only by artists sculpting in wood.

18th-century door arch shaped with a hand adz and then planed

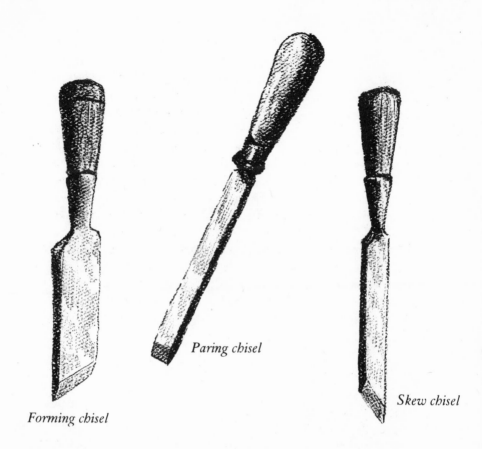

Paring chisel

Skew chisel

Forming chisel

CHISELS

Another group of basic carpenters' tools, which were used in the Stone Age and developed through the Copper, Bronze, and early Iron ages, are the chisels. At the time America became settled the tool had evolved into two basic types, forming, or firming, chisels and paring chisels.

Some **forming chisels** were socketed, by having the end of the blade shaped into a cone into which the handle was fitted; others were shaped with a tang and a wide shoulder, the tang being driven into the end of the handle. Forming chisels generally had a bezeled edge. They cut by being hit with a wooden mallet which drove the edge into the wood.

Paring chisels were usually tanged, with a slim handle. They almost always had knife edges and were never hit with a mallet, but pushed to smooth a rough cut made by the forming chisel; they were used as finishing tools.

There were a number of types of forming chisels, each type being made in different sizes by the local blacksmith before 1840 or 1850, and by the growing tool industry after 1850.

Almost all carpenters used several **flat chisels** with flat blades, bezeled on the end to form an edge, and in sizes from a quarter-inch to a two-inch width across the edge. These were used mainly to carve mortises after a series of holes had been bored, and to split off excess wood when forming tenons. **Skew chisels,** named for the angled edge, were used to clean out the corners of mortises.

When soft wood such as pine or spruce was used for mortised house timbers the carpenter used a special **mortising chisel**. This tool had a blade almost square in section with a deeply bezeled edge. It was used to start a mortise without the usual boring of holes, thus saving a great deal of time.

Door mortise cut with forming and paring chisels

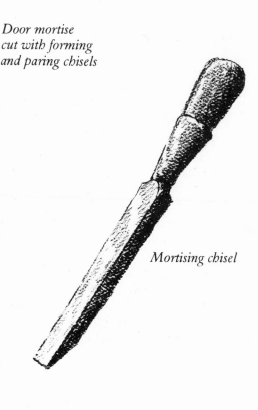

Mortising chisel

House timber mortise of the sort cut with a mortising chisel

Pennsylvania housewrights and millwrights used a couple of special forming chisels developed in northern Europe but apparently never widely adopted by the British nor by the British colonists in America. These were the mortising ax and the **twibil**, the latter being used widely in Scandinavia. Both were used only for mortising.

Shaped like a chisel with its socket parallel to the edge, the mortising ax was held in position by its wooden handle and pounded with a mallet at the intersection of socket and blade. This tool was used in the Pennsylvania backwoods for barn building until the 1890s.

The twibil, which may be roughly translated into "two edges" (the old English pruning tool, a hooked knife on a long handle, was named a "billhook," or "edged hook") also had its handle socket at right angles to the double-ended chisel blade. One edge of the twibil paralleled the axis of the handle as with an ax, the other being sharpened perpendicular to the handle as with an adz. Twibils were sometimes swung like an ax or adz, but usually were pounded with a mallet on the long thick socket.

Mortising ax

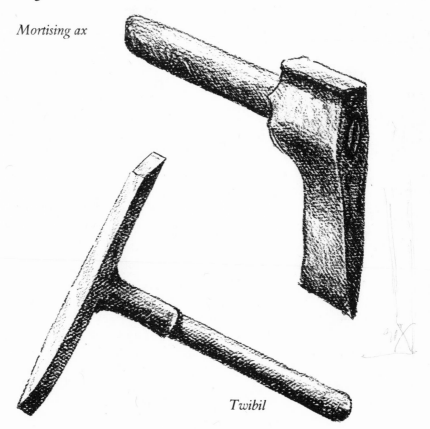

Twibil

Assembled house frame of mortised timbers

Corner chisels were used almost exclusively to finish the corners of mortises that were formed with a combination of auger holes and chisel cuts. This form of chisel is now as obsolete as the hand-adz.

Some house carpenters carried a few **gouges** in their tool chest. This tool, represented today only by wood-turning chisels and carving tools, had really very little use from a carpenter. A few used it to rough out large cavities, the curved edge cutting more easily than the straight edge, as with the jack plane blade. Of course, the cuts of a gouge were usually dressed, or finished, with a flat chisel.

While all the forming chisels were struck by a mallet to make a cut, the paring chisels were all pushed with the hand. As a consequence paring chisel handles were narrow and graceful, easily grasped. The blades were narrow sections of knife blades sharpened on the end instead of the long edge. In effect they were identical in basic design and the techniques of their use to the woodcarving tools used by the cabinetmaker.

When interior door locks were invented after the Civil War, their installation required a deep mortise in the edge of the door. As with other mortises, the lock mortise was started with a series of holes bored with brace and bit, then finished with a chisel. Many carpenters, however, found that chiseling was much easier with a special paring chisel called a **gooseneck**, shaped so that it could be inserted into the lock cavity for cutting corners which could not be seen. Modern door locks require only a hole quickly bored with a power drill.

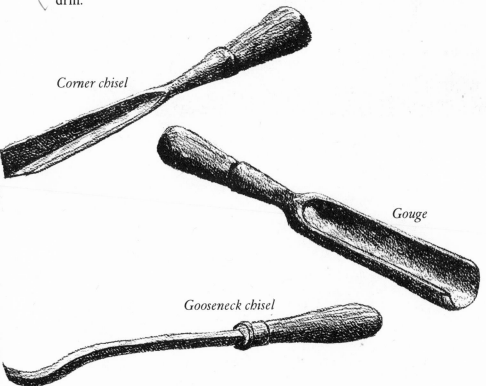

Corner chisel

Gouge

Gooseneck chisel

House timber mortise finished with a corner chisel

Door lock cavity hollowed with a gooseneck chisel

The largest chisel sometimes used by carpenters was called a **slick**. Actually it was a giant paring chisel with a blade up to three or four inches wide and from eighteen to twenty-four inches long. A wooden handle made the overall length around three feet. The slick was used to trim framing timbers or floors. It was held in both hands, or placed against one shoulder, and pushed to delicately pare large uneven surfaces which were hard to work with a plane.

Slick

Floor smoothed with a slick after laying

AWLS

While building houses in Colonial times carpenters occasionally had to nail thin lumber such as molding or furring strips, pieces of stuff so thin that the incidence of splitting when nailed could create a problem. The problem was intensified in a day when lumber was handsawn and planed by hand. The **brad awl**, probably the oldest of the boring tools, solved this problem. Consisting of a small steel rod with a chisel point, mounted in a chisel-type handle, it was rotated back and forth in thin stuff until a small hole was bored. A nail could then be driven through without splitting the stuff. The clockwise and counterclockwise motion used with the brad awl is a primitive technique exactly duplicating the motion of a bone awl used to pierce leather in the Stone Age or that of a Stone Age pump drill, indicating that the brad awl is at least as old in concept as the nail. It was used by American carpenters, especially in Pennsylvania, until the advent of small wire nails late in the nineteenth century.

Brad awl

AUGER AND BRACE AND BIT

Another ancient boring tool, the auger, cut large holes in heavy timber by being rotated in one direction, usually clockwise, by means of a 15-to-18-inch handle mounted perpendicular to the bit. Throughout the early settlement of America until the end of the eighteenth century the bit used was called a **spoon bit** because it resembled a spoon with sharp cutting edges. The spoon bit cut only when enough downward pressure was applied to make its point dig into the wood. A variation of the spoon bit was the **shell bit**, shaped like a gouge and used to remove a round plug from a board, creating a hole. Both spoon and shell augers required a great deal of downward pressure, and both had to be removed fairly frequently to clear wood chips from the hold. The **pod bit**, the point of which was swelled and slightly twisted, was often used to start a hole which was finished with a shell bit.

An improvement on these bits came with the cutting auger, a **center bit**, found from about the sixteenth century in most carpenters' tool kits. It had a point which held the bit in one spot, and it had both a sharp scriber which cut through the outline of the intended hole and a small blade which cut out the wood encircled by the scriber. It was used only to bore crossgrain, never end grain. As with spoon and shell bits, however, removal of wood chips from the hole was a problem. This was solved to some extent by the ancient invention of the **twist bit**, still in use at the time of early American colonies. The twist bit was equipped with the point of a center bit. Its advantage lay in the twist it was given above the point which served to push out the chips.

Then, around the year 1809, most of the ancient problems of boring a straight, accurate hole through a board were solved with the development of the **Scotch bit**, which appeared simultaneously in America and Europe. The Scotch bit was essentially a twist bit with cutting scribers and cutting blades on its end. Instead of having a sharp point, though, it had a screw point which drew the scribers and cutting edges into the wood, all the while guiding the direction of the hole.

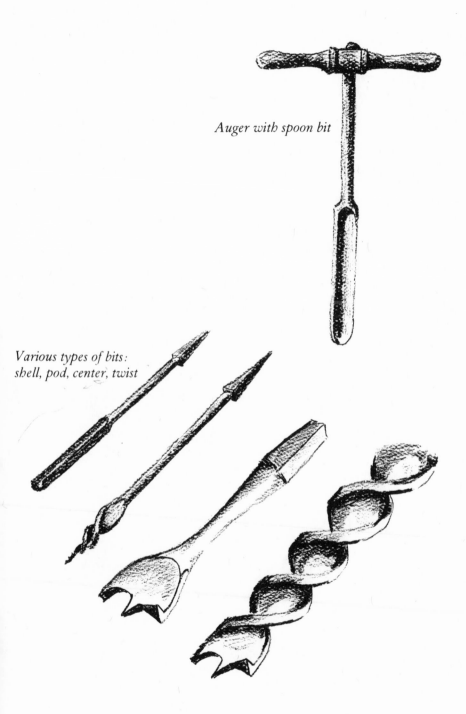

Auger with spoon bit

Various types of bits:
shell, pod, center, twist

The Scotch bit made it much easier to bore holes, and it was almost universally adopted by woodworkers from the time it first appeared.

Coincident with the discovery and colonization of America was the European invention, or perhaps adaptation from the Orient, of the **carpenter's brace**. Braces were at first known as piercers and apparently were used to make an initial hole which was later reamed to larger size with a spoon auger. The brace, however, became an essential part of the carpenter's and housewright's tool chests by the eighteenth century. Because it allowed a continuous clockwise rotation to the bit, instead of the broken rhythm of the auger, and because it was easier to apply downward pressure, brace and bit soon replaced the auger for boring holes up to three-quarters of an inch in diameter.

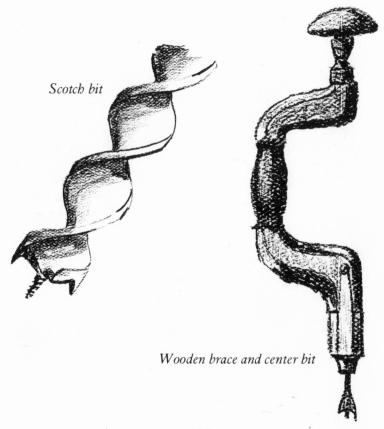

Scotch bit

Wooden brace and center bit

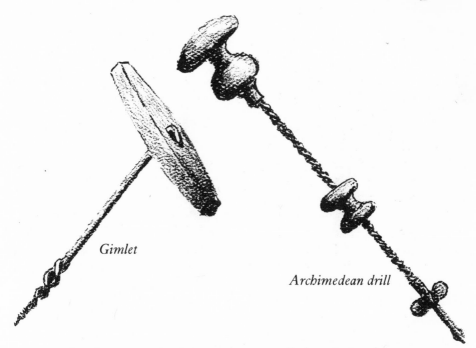

Gimlet

Archimedean drill

Another type of miniature auger, called a **gimlet**, has survived from ancient times but enjoyed something of a revival when nails and brads for fastening hinges and other house hardware were replaced by wood screws early in the nineteenth century. The gimlet was, and to some extent still is, used to quickly bore a small, shallow hole into which the wood screw is inserted. Gimlets are still available in most hardware stores, where augers can no longer be found.

Carpenters depended on the gimlet and brace and bit until after World War II when the advent of power drills began the certain obsolesence of most hand tools. Since that time the electric drill has practically replaced all of the augers and braces and bits. Interestingly, the bit developed by modern engineers for high-speed boring with an electric drill closely resembles the old center bit which preceded the improved Scotch bit.

A few modern carpenters still use a reciprocating or **Archimedean drill**, generally known as a push drill, for boring small holes to receive wood screws. This invention appeared about 1900 in America and Europe, but never wholly replaced the gimlet. It is still available today from mail order houses and tool suppliers.

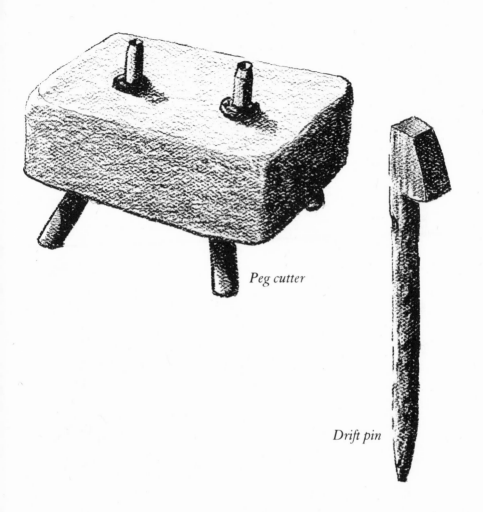

Peg cutter

Drift pin

PEG CUTTER AND DRIFT PIN

Since the principal reason for boring holes in making a house was to insert tree nails or trunnels (wooden pegs that secured joints by pinning mortises and tenons together in the framing timbers), it was only logical that the housewright should also devise an easy means for making pegs. Often trunnels were square, split out of a block of straight-grained oak or pine and forced into the round hole. Round pegs were also used, however, and these were made by driving a square peg through a tubular peg cutter mounted on a block of wood that had a hole directly underneath the cutter so that finished pegs could fall through to the ground as they were shaped.

The house and barn frames of pre-1840 were fitted together on the ground and the measurements checked before being raised to final position and fitted to other timbers to form the finished frame. Old housewrights used an obscure and easily overlooked but nonetheless important tool for checking the fit of joints. It was called a drift pin and was whittled roughly out of scrap lumber. Drift pins were inserted in the trunnel holes and were removed by tapping the hook-shaped head, after which the trunnel was pounded in place.

MISCELLANEOUS TOOLS

Besides the expected hammers and saws and planes and chisels, the eighteenth- and nineteenth-century carpenters used a host of small miscellaneous tools, most of which had become antiquities by the latter half of the twentieth century. All these small helpers reflected the self-reliant days before the retail availability of a number of conveniences so common in the modern age of obsolesence that they are seldom given a thought. For instance, although pencils have been available since the time of the French explorations of the Mississippi Valley in the seventeenth century, pencils were not readily available to the early English, Dutch, Spanish, and Swedish colonists on the eastern seaboard of America. But since carpenters must have some means of marking boards, the early carpenters each had a **scriber,** no more than a pointed metal rod, often of tempered steel, which was found in a number of forms. Some were merely small rods sharpened on both ends, some were handled, and some had a 90° bend on one of the sharpened ends. Those with the bend could be used to mark a board to be ripped, by holding the scriber with the bent point extending from the hand and pulling it down the length of the boards, the knuckle of the forefinger or the end of the thumb being used as a fence to control the path of the point. Actually a scriber is a better marking device than a pencil, for it produces a hairline to guide the saw while a pencil line varies in width up to a 1/16th of an inch, too much tolerance for tight joining.

Scriber, showing handhold

A more accurate, and more specialized, marking device was the **marking gauge**. Apparently this ubiquitous tool is a relic of Roman days at least and was common in its home-made form in Colonial America, and common in its factory-made form from about 1840 until World War I. It consists of a straight board, some 6 or 7 inches long, marked on one side into a rule. This stock has a small metal scriber protruding from one end, and the square stock is inserted through a small fence board with a wooden thumbscrew in one edge that is tightened to hold stock and board. To use it the carpenter sets the fence a specified distance from the scriber, tightens the

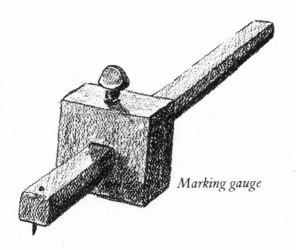

Marking gauge

screw, places the fence along the edge of the board to be sawed, and pulls the gauge the length of the board.

Another ancient marking device is the **chalk line,** simple in concept and accurate in application. It consists of a long cord which is coated with chalk or charcoal dust, held taut on a board or log to be marked with a straight line and snapped so that the dust is transferred by the line to the timber. Often the line was carried in a loosely wound ball and the dust in a small tin or wooden box. Eighteenth-century carpenters, however, sometimes made a very pretty container to which was attached a small reel, with crank, for the cord. Sometimes the ingenuity of the preindustrial, preengineering age was reflected in a chalk line reel and box made from a cow horn.

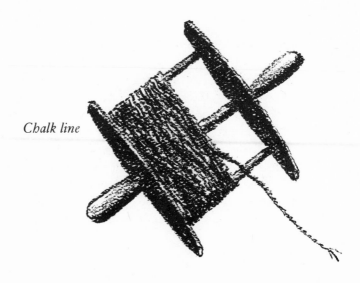

Chalk line

Levels and squares were fully as important to the early American carpenter as to his late-twentieth-century mechanized descendant. Before 1840, though, each carpenter usually made his own, out of wood, according to basic logic and ancient design. Until about the time of the Civil War the level was no more than a straight hardwood board with a triangular, semicircular, or straight projection extending from one edge. A cord with a small plumb (or lead weight, from the Latin *plumba* meaning lead) was attached to the center of the projection. When a beam or floor was leveled, the cord hung straight so that the plumb bob matched a mark in the center of the level. **Spirit levels,** boards equipped with a small glass tube of alcohol or oil and a tiny air bubble which floated in the center of the tube to show levelness, were not readily adopted until after the Civil War. The principle of spirit levels, however, had been long established by an old expedient used for leveling beams whenever a plumb level was unavailable; frontier housewrights often used a small glass or saucer of water to check the level of a joist or sill.

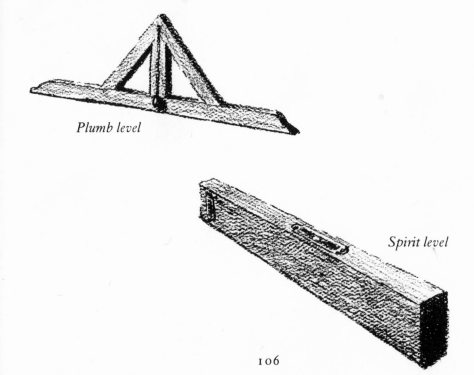

Plumb level

Spirit level

But even after a sill was leveled a house could still be crooked unless all studs and joists were foursquare, and this could only be checked by a large framing square consisting of two boards about two and three feet long respectively, fastened together to form a perfect right angle. The carpenter who did not have a square to check the one he was making proceeded confidently by applying the ancient Greek geometric axiom of 3-4-5: if two lines of three and four inches intersect at a right angle, then the distance between the ends away from the angle will be exactly five inches. **Marking squares,** used for marking the end of a board to be sawn, had legs of about six and twelve inches.

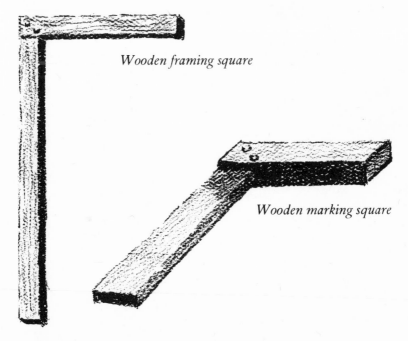

Wooden framing square

Wooden marking square

Metal framing squares, which quickly replaced the homemade wooden variety, were first made in New England about 1840 by a blacksmith who was so successful that he founded a new industry. His development has not yet been displaced, and its only improvement has been the addition of stamped rules along each leg of the square.

An indispensable tool to the carpenter before 1900 was the saw set, a tool now so obsolete that most home craftsmen cannot even define its function. It was used to set saw teeth after a saw had been sharpened with a small file, a job the old-time carpenter would seldom trust to another. The saw set was, in principle, a metal bar with slits that fitted over the saw teeth to bend them outward and create the set which kept a saw from binding. Most old saw sets had a series of slits to fit the saws of various degrees of coarseness which were used by any one carpenter. The simple saw set, which had been around since Roman days and required a good eye to put exactly the same set on each tooth, was replaced in the late nineteenth century by a more complicated mass-produced hand tool which could be adjusted to different-sized teeth and different angles of setting.

Setting and sharpening was especially important maintenance for saws used on large construction jobs between the two world wars, when carpenters were used mainly to build wooden forms for poured concrete members of a building. On these jobs the foreman often assigned the oldest or least agile of his carpenters to spend full time sharpening and setting the saws used by the whole crew. Such constant maintenance was made necessary by the concrete residue on reused boards which dulled saws quickly.

Saw set

After World War II few professional carpenters used a handsaw enough to require setting, and when they did they generally took their saws to special shops for sharpening and setting.

A **screwdriver** seems to be the simplest of tools and it is found in every modern carpenter's toolbox, ready to use for installing locks and hinges on doors, putting handles and locks on modern sash windows and for a number of other uses within the woodworker's milieu. Actually, in terms of history, the screwdriver is a relatively recent tool for the carpenter. Few wood screws were used in common carpentry before 1800 and before that date the carpenter would seldom have had use for a screwdriver in his day-by-day work.

Another simple tool new to the trade is the **nail set**, a small tapered punch used to drive the heads of wire finishing nails below the surface of a board without marring the board with hammer marks. Nail sets did not come into general use until the 1880s and '90s when wire nails were first introduced from England.

Woodworker's screwdriver

Nail punch

Master carpenters also used a few rather more sophisticated tools than have been mentioned here. Since he shared them with the cabinetmaker, however, in that area of dim differentiation between master carpentry and cabinetry, these few tools, such as the bevel and miter box, will be examined among the tools of the cabinetmaker's bench.

MODERN TOOLS

A person unfamiliar with woodworking techniques might be amazed by the variety of hand tools still offered in hardware stores; in reality, however, there has been little variety at all since 1900. There are still hammers, but of only a few standard weights, all of them with adz-eyes and many with steel or fiberglass handles, although many hammers are still sold with good, honest hickory handles.

Hatchets with bezeled edges are often available as is the shingling hatchet with a knife edge. The form and weight of modern hatchets are quite standardized, however, regardless of the manufacturer, all of them conforming to the needs of mass production rather than the tastes and needs of the user.

Handsaws may be bought, though many hardware clerks no longer understand the difference between a crosscut saw and a ripsaw. Coping saws have not yet been displaced entirely by the machine and keyhole saws still find a narrow market in most communities.

Forming chisels are still a necessary carpenter's tool, and are available in various widths in most hardware stores. The same applies to wood bits and braces which, while not commonly displayed, are still manufactured and distributed in most areas.

It is the family of planes which probably has suffered the most attrition in the industrial age: a matter of beauty and the beast. European toolmakers of today still manufacture a fair variety of planes with wooden stocks, although this type was abandoned in America in the latter part of the nineteenth century and replaced by metal-stocked planes.

Today in America there is only one type of plane in three or four sizes, all of which are smoothing planes, the largest being miscalled a jack plane. The rest—the molding planes and rabbeting planes and moving fillisters—are lost. And gone with them are the interesting shapes, the ingenious designs, and the beauty of wood stocks polished with work, reflecting the thousands of strokes by each plane in making hundreds of pieces of molding.

Hardware stores now specialize in all manner of power tools: saws and drills and shimmying sanders. No professional carpenter since the 1950s would even consider building a house with handsaws, and after this time many houses were built with no crown mold, no base mold, and often no door or window mold. Those houses with some indication of former elegance would certainly have all molding supplied by a mill, not planned on the house site. More and more houses, though thankfully not the majority, are being put together with staple guns and adhesives, and this makes even the basic claw hammer somewhat obsolete.

The trouble is that, since about 1900, many buildings, those that will serve as monuments to the workmen of the era, are far more likely to be built with steel and prestressed concrete than with wood. Wood is widely used in residences, but more and more metal window units are being used, more and more carpet floors are being installed on bare concrete slabs, and more and more all-steel buildings are being built, including some residences, a far cry from houses in the Colonial period which were often built entirely of wood without even the use of a single handwrought iron nail.

The affinity of man with wood, however, seems likely to continue. The best residential architecture still calls for wood in floors, decorative molding, and paneled walls. When it is possible to obtain, and can be afforded, many Americans still desire hand craftsmanship, perhaps because they feel nostalgic for the old days or perhaps because wood, grown by God and worked by the hands of a man instead of a machine, offers a cozy comfort that will never be displaced by concrete and steel.

3

THE FURNITURE MAKER AND THE CABINETMAKER

American houses from the sixteenth century to the middle of the nineteenth century were well built, but the furniture which filled these old houses was built with remarkable precision. The joints of much of the old handmade furniture are so tight as to be almost invisible. The engineering and design of handmade chairs and beds is so well related to the function of the pieces that some furniture three hundred years old still has more durability than modern furniture mass-produced and assembled with modern techniques.

Perhaps this was achieved because the older handcraftsman took such pains with the joints and design of his pieces. Loving care and professional pride, reflected in each individual stroke of his work, combined with an encyclopedic knowledge of his material, were factors beyond the capability of a machine or a machine operator.

Furniture and its component parts normally display far greater exercises in form than do houses. It was not until late in the nineteenth century that houses began to be trimmed with all manner of gingerbread decorations, but furniture from Roman times onward, and certainly during the Colonial, Federalist, and Victorian ages in America, was capable of an infinite variety of ovals, circles, molding sections, and fretwork in many different styles. Whereas the carpenter dealt mainly with straight lines, the cabinetmaker was called upon to work in straight lines plus sculptured lines of every conceivable shape. Whereas the carpenter depended on plumb line and framing square, the cabinetmaker resorted to ogee curve and thumbnail mold. And whereas the carpenter performed his relatively heavy tasks at the housesite, the cabinetmaker was dependent on his shop, complete with bench, vise, turning lathe, and all the tools needed to provide decoration, elegance, and unseen joints.

Since both craftsmen worked in wood, the tools of each were identical in principle although they varied greatly in form. The cabinetmaker needed delicate tools for delicate work, and he often made his own for special jobs. His work, too, was more varied than that of the carpenter, and while a master carpenter might own sixty or seventy individual tools, the master cabinetmaker sometimes required as many as two hundred, plus the mechanical help of lathes, screw vises, and treadle saws. Often, because of the need for tools which could not easily be transported to the housesite, the carpenter ordered special panelling, mantelpieces, stair banisters, or gingerbread decorations from the cabinetmaker, and installed them himself.

A look at the cabinetmaker's shop will soon show the difference between his tools and those of his colleague, the carpenter.

Late 17th-century chest

THE WORKBENCH

Because the cabinetmaker worked with relatively small pieces of wood he needed a solidly built bench, of a height comfortable to stand at and equipped with all the devices needed to provide extra hands to hold work steady while he planed, sawed, or carved.

In early Colonial days in America the workbench was much simpler than after 1850 and it generally followed the English pattern which, for some unknown reason, never adopted one or two ingenious refinements found on the Continental bench. French and German cabinetmakers, however, undoubtedly built the type of bench used in their apprenticeships on the Continent, and a Continental bench is illustrated for that reason. Usually in shops of any size, such as those found in the early eighteenth century in New York, Philadelphia, Boston, or Charleston, each journeyman cabinetmaker, as well as the master, would build his own bench to suit his own techniques and certainly to suit his size. For a man of average height the top of the workbench would be about thirty-six inches from the floor, but there was no stan-

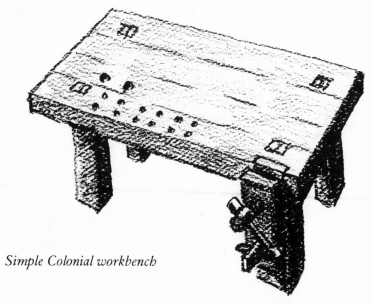

Simple Colonial workbench

An 18th-century handmade spoon chest

dard. The important thing was to have the bench high enough to work on without strain, and it usually stood so that its top would strike midway between the worker's elbow and wrist. At this height he could work with plane or saw or carving tool all day long with a minimum of back-bending or arm-lifting.

There were a number of integral auxiliary holding devices on all cabinetmakers' benches, both in the Colonial period and after the beginning of the industrial age, and the sturdy two- to four-inch tops of the old benches were drilled with a number of holes to accommodate many of these concomitant members.

HOLDING DEVICES

Perhaps the simplest of the holding tools was the stop, in use since Roman days. In its older form and until well into the nineteenth century it consisted of a small iron rod with a flat rectangular head, set at a slight angle and equipped with teeth on the higher edge. This was dropped into one of a series of small holes in the bench top, the lower edge of its head resting on the top, the teeth elevated to grip the edge of a board laid on the bench to be planed. Usually several stops were used for each board, one or two on the end and three or

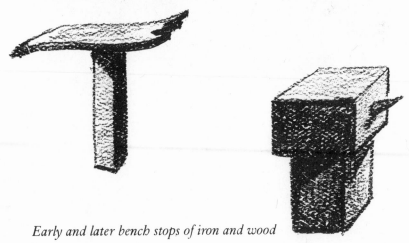

Early and later bench stops of iron and wood

An early 19th-century bookcase and chest of drawers

four along the edge to hold it securely while the plane smoothed its flat surface. After about 1820, when an improved bench design with a built-in end vise became popular, the ancient metal stop was discarded in favor of small square wooden stops which protruded about a quarter-inch from square holes in the bench top. Boards being planed were clamped tight between two wooden stops, one on the bench, the other inserted in the end vise.

For planing the edges of boards, the seventeenth- and eighteenth-century English cabinetmaker often used what was generally called a **bench clamp**, but was sometimes referred to as a holdfast. It consisted in its early form of a flat board nailed to the bench top with a V notch cut into its end. Boards to be planed along the edge had one end jambed into this notch, which held the stuff steady while being planed. Nineteenth-century bench clamps were sometimes more elaborate, consisting of two small boards fastened to the top at angles to each other, and two wedges which were tapped into the angle with the stuff being worked held tight between them. Continental benches usually were equipped with a clamp made from an angled tree limb which was fastened on the front edge of the bench. The stuff was jambed into this angle, its length supported by removable pegs in holes in the bench legs.

Two types of bench clamps, the front clamp being typical of Continental workbenches

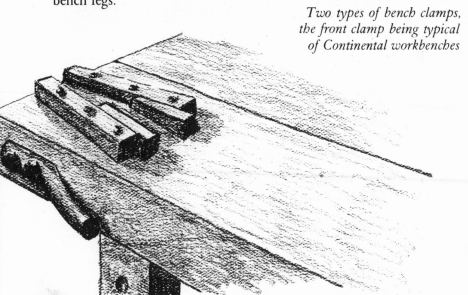

Hand-planed boards in an early 18th-century cabinet

Certainly one of the most ingenious holding devices used by ancient and Colonial craftsmen alike, and one which unaccountably has disappeared, is the **metal holdfast**, a roughly inverted L-shaped tool which was jambed into a hole in the bench top to securely hold boards being carved or planed. The efficiency of this simple tool is quite amazing, almost as amazing as the fact that it has been wholly displaced by the vise since 1900.

Another simple tool which in America has joined the holdfast in the limbo of obsolete tools is the **side rest**, used to hold stuff being sawed on the bench top, raising the board an inch or so to prevent the saw's scarring the bench top. Side rests were usually made from a thick, solid board, sawed to leave projections on opposite sides, one to fit against the front of the bench top, the other to serve as a stop for the board being sawed. Sometimes side rests were narrow and used in pairs; but some were wide and used alone.

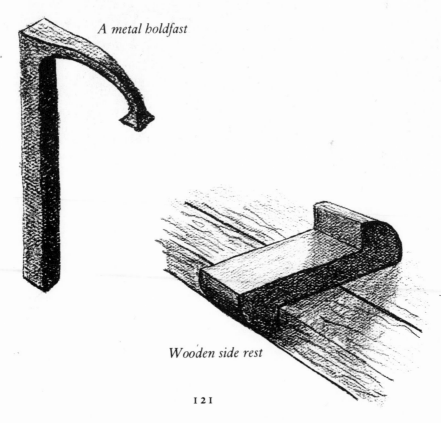

A metal holdfast

Wooden side rest

Detail of an 18th-century walnut secretary

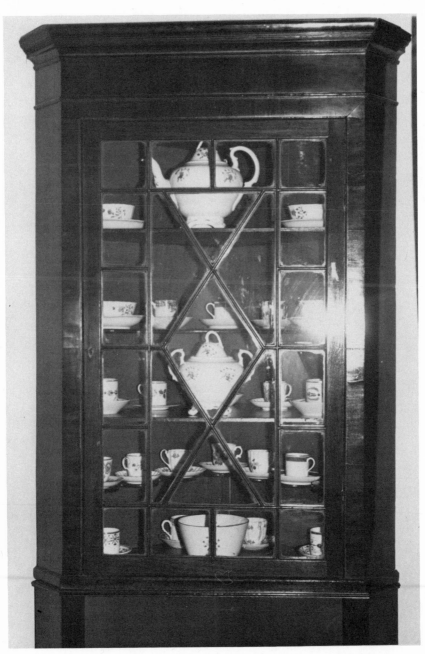

Corner china cabinet

The most complicated and mechanical of workbench auxiliary tools were the **screw vises**, generally unknown except in isolated cases until the last quarter of the seventeenth century. From about 1675 until 1820 only the side vise was used. This consisted of merely two stout wide boards, one, the backboard, fastened vertically and immovably to the front of the bench top and a bench leg, with a threaded hole to receive the screw. The other, the front, was pierced in two places, one hole for the screw and another, usually rectangular hole, to fit over a smaller board which was mortised into the back. This smaller board was bored with one-inch holes about two inches apart through which a dowel was placed to keep the two boards parallel. Vise screws before about 1870 were usually of hardwood, two inches or more in diameter and made in the cabinetmaker's shop. Such a vise could tightly hold any board being planed or molded on its edge, and generally was used to clamp pieces, such as the front and side of drawers, while they were being assembled. Later, the end vise, already described, was added to the bench.

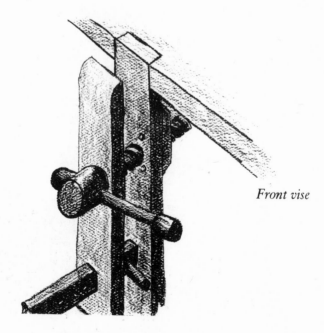

Front vise

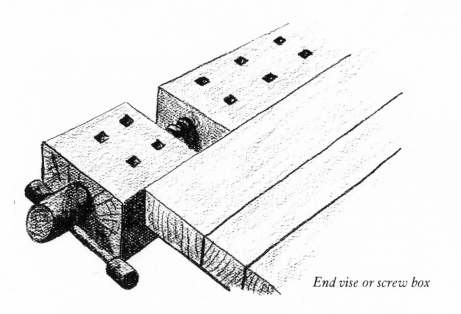

End vise or screw box

A pegged mortise and tenon joint

By the beginning of the nineteenth century workbenches became more elaborate, some having a sort of well across the back in which to place tools handily, others with drawers for tools and perhaps a shelf bored with a series of holes in which chisels and bits could be held.

In addition to the workbench the shop generally had shelves on its walls, convenient to the bench, on which planes were set, saws were hung, and other tools were kept for ready access.

Though the cabinetmaker did not face the monumental task of hewing sixteen-inch house sills, he did resort often to the ancient hewing tools for roughing pieces of furniture into shape. For this job he used essentially the same hewing tools employed by the carpenter, but in lighter versions for lighter work. He eschewed the heavy broadax for the light hatchet, razor sharp. He shared the hand adz with the carpenter, but generally used a lighter version. For shaping chair seats in Windsor chairs he employed a foot gouge-adz, but it was quite light and shaped especially to fit his needs.

Later bench of Continental design

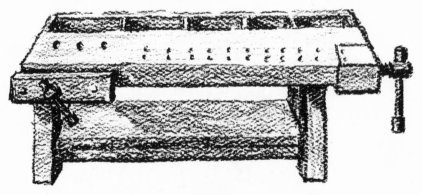

Cabinetmaker's light hand adz, also known as a cooper's adz

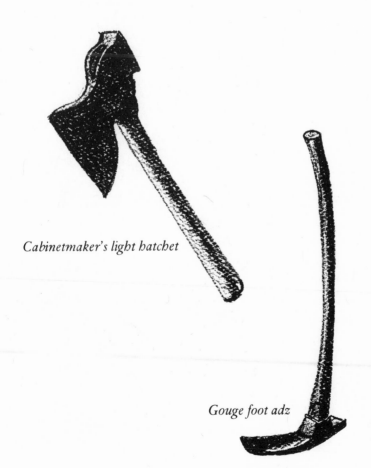

Cabinetmaker's light hatchet

Gouge foot adz

Detail of an early 19th-century pine table

Chair arms which were probably roughhewn with a hatchet before final shaping and smoothing

Windsor chair seat hollowed with a gouge foot adz

To a cabinetmaker the glue pot and clamp were far more important than trunnel and nail. Consequently, the cabinetmaker seldom used the claw hammer except to make packing crates. Even then, his hammer was likely to be of lighter weight than that of the carpenter.

Since nails were not often used, the cabinetmaker used a light eight-ounce Warrington, or London Pattern, hammer for occasional use in driving brads or striking special chisels not equipped with wooden handles. The Warrington hammer resembles a cross peen machinist's hammer. The date of its design and the source of its name are unknown.

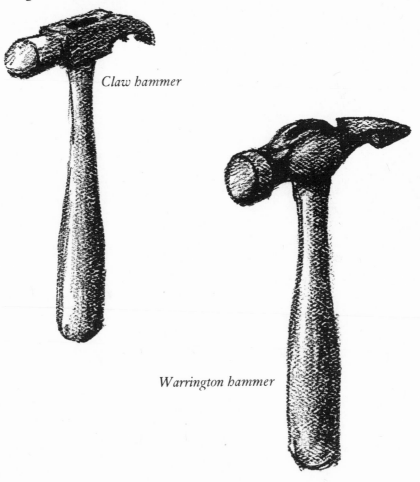

Claw hammer

Warrington hammer

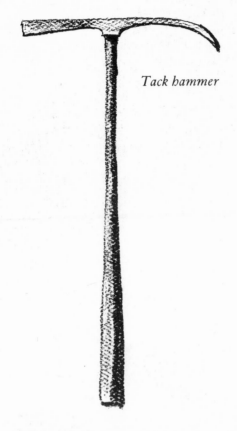

Tack hammer

The American cabinetmaker, however, was frequently called upon to be upholsterer as well as woodworker, and upholstering did require the use of tacks and small brads. Also, some of the delicate molding was often fastened to a piece of furniture with small iron brads. To drive tacks and brads the cabinetmaker kept a graceful, light tack hammer with an elongated face and short claws fashioned on the end of a graceful poll.

The other striking tools used in a cabinet shop were wooden mallets and mauls in various sizes and weights. These were used to strike chisels, drive mortises and tenons tightly together, and to drive the pins which secured mortise and tenon on large pieces.

Detail of secretary showing pegged mortise and tenon and molding held on by small nails

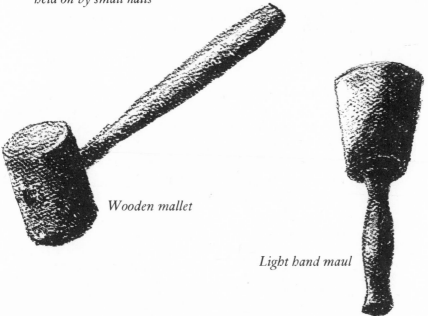

Wooden mallet

Light hand maul

Most of the old cabinetmakers' saws were framed, although they had need for several varieties of backsaws and small, narrow handsaws.

Handsaws, exactly like those used by the carpenter, were referred to as panel saws by the cabinetmaker, and used for rough cutting of the stuff which later would be mitered, planed, tenoned, and grooved.

In addition, however, the more precise craftsman who made furniture required a number of special saws which were common from the time of America's first settlements until about 1900 when the demands of a fast-growing population could no longer be supplied with hand labor.

For the delicate, precise labor of forming dovetails the craftsman used the dovetail saw, a specialized instrument with either a turned handle into which the tang of the saw was inserted or an open handle of the type used on eighteenth-century English handsaws. The blade was of thin steel about two inches wide, stiffened with a narrow plate of iron or brass which was doubled over the back edge of the saw blade to keep the blade from bending while sawing was in progress, a minor catastrophe not to be tolerated in a close dovetail joint. Most cabinetmakers also used a saw with no set to the teeth and a handle offset to the blade. This was used most frequently to cut off mortise pins close to the wood without scarring it.

Tenon saw

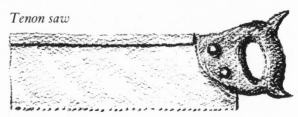

Dovetail saw

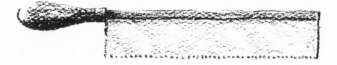

Larger backsaws up to five inches wide and twenty inches long, and usually with an open handle, were used in cutting miters for molding and picture frames, and in forming tenons. This saw, identical to the master carpenter's backsaw, was often referred to in cabinet shops as a tenon saw.

18th-century secretary with tenoned frame and doors

Hand-cut dovetail joint in a pine chest

Very small openings, such as keyholes, were formed with a keyhole saw, often in conjunction with the auger. Certainly the hole from which the sawing started had to be bored, and the saw then cut out the proper shape. Keyhole saws had very narrow but thick blades, with fine toothing and fine set, often with a round, turned handle into which the tang of the saw was inserted.

Keyhole saw

Keyhole formed with a keyhole saw

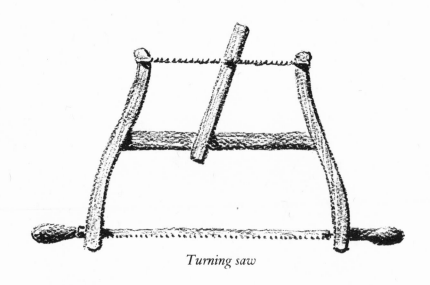

Turning saw

Turning saws, or coping saws, or in the frequent misnomer of modern parlance, jigsaws, were frame saws, sometimes known as bow saws. Until 1900 the frame was usually of wood, the thin, narrow blade being stretched just as in the bucksaw used by the carpenter. Turning saw blades frequently had the teeth set a little coarser than other saws. It also had turned handles, protruding from both sides of the frame. In each handle was a rod, hooked on its end, which was inserted into a hole in the end of the blade. The turning saw frame, of course, was smaller than that of the bucksaw. By virtue of the handles, the blade could be turned to various angles to make it more versatile in cutting both interior and exterior shapes on work held in the vise. After 1900 the turning saw was reduced in size and held in a metal frame. Turning saws, as opposed to other types, usually cut with a pulling motion, which eliminated the chance of the thin blade buckling while in the kerf.

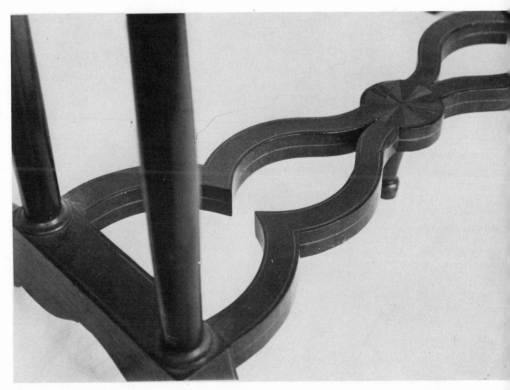

Decorative table stretcher formed with a turning saw

Stair tread housing cut with a rabbet saw

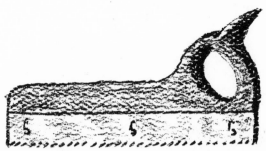

Rabbet saw

Rabbet saws are ancient tools that were generally made by the cabinetmaker until a design was patented in America in the middle of the nineteenth century. They were used mainly to cut the mortises in the walls of bookcases into which the shelves fitted, or, when the cabinetmaker became involved in the special woodwork of a fine house, to cut the hidden housing into which stair treads were fitted.

The traditions of handwork and the importance of training the hand and eye of an apprentice persisted longer among cabinetmakers than in other trades. Even when mechanization of a sort appeared in the cabinet shop it had the nature of a missing link of industrial revolution. The first major machine after the lathe, which already had a long pedigree by the seventeenth century, was probably the bench jigsaw, used to replace the turning saw, the foot replacing the hand as a source of power. On this simple machine, which apparently appeared during the late eighteenth or early nineteenth century, the saw blade was fastened upright in a frame between wooden springs and a foot treadle. Its blade passed through a hole in a stationary table set carefully at right angles to the saw blade. Sawing commenced when a piece of stuff was placed next to the blade and the foot treadle pushed down to start the cut, the blade being returned by spring action. This simple machine must have been beloved by apprentices who could use it to guarantee that fretwork, and other shapes formerly produced with the turning saw, was always cut square to the flat surface of the stuff (it sometimes took years of experience to learn how to keep the turning saw square).

It is probable that some of the intricate gingerbread decoration found in the Hudson Valley architecture of the 1840s and 1850s was produced by the treadle jigsaw after it had been adopted by lumber suppliers in the very beginning years of this industry. After the Civil War, when gingerbread became *de rigueur* on the fashionable houses of the times, it must have been produced by the newly established, specialized planing mills which also furnished standard molding and flooring. By this time, however, the treadle jigsaw had been converted from foot power (which required knee grease instead of elbow grease) to steam power.

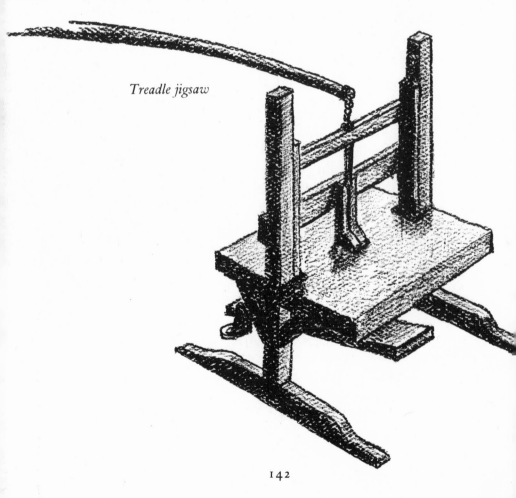

Treadle jigsaw

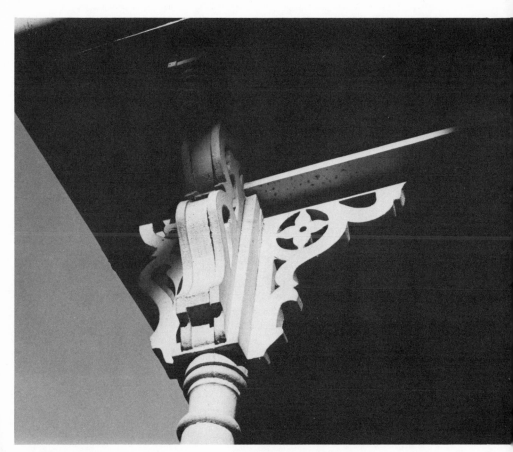

Gingerbread porch decoration probably cut with a treadle jigsaw

An important adjunct to saws was the miter box, a jig used on the bench to cut accurate angles across a board or piece of molding. Although very precise miter boxes made of metal and adjustable from 1° to 179° have been available from industry since about 1900, carpenters and cabinetmakers in the 1970s occasionally continue to make simple miter boxes from three thick but relatively narrow boards. A channel is formed by nailing two boards, usually 2″ x 4″ or 2″ x 6″, set on edge, to the edges of a third board set flat. Saw cuts at 45°, 90°, or any required angle are then carefully cut through the two boards set on edge until the saw reaches the bottom board. These kerfs guide the saw when cutting a board placed in the channel. A slightly more accurate miter box was made in the old days by gluing or screwing a board previously cut to the intended angle to the top step of a stepped block. Boards to be cut were held on the bottom step and the saw guided by the angled board, cutting through both the top step and the board being mitered. The larger support surface on this latter type guaranteed closer accuracy for cuts. In extremely accurate work miters were always finished by planing in a miter shooting block.

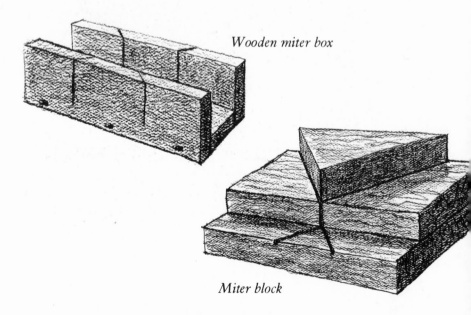

Wooden miter box

Miter block

China cabinet with mitered molding

In the heyday of handcraftsmanship in wood, each cabinet-maker had a love affair with his planes. It was this tool above all others which allowed him the close precision of the professional, which provided the necessary duplication of a molding used several times on a piece of furniture, and which expressed his originality, his sense of design, and the artistry of his stroke. Planing was referred to as shooting, and the master cabinetmaker learned to shoot as accurately with his planes as any fabled backwoodsman with his Kentucky rifle.

Besides the common planes used by the carpenter, the cabinetmaker made himself special planes whenever needed, each designed for a different use and varying in size from two inches to two feet in length. In connection with his planing he also made himself a number of auxiliary pieces of equipment such as scrapers, shooting or molding boards, and miter shooting blocks.

The planes were mostly for carving moldings on the edges of tables and desk and dresser tops, a wide variety of types being needed to fit the various convolutions found on these tops during the eighteenth and early nineteenth centuries. Most contradicted the name "plane," for most had soles designed to cut concave or convex molding and were shaped in an arc rather than a flat plane. Those which cut moldings on a concave edge were generally called compass planes as they traveled an inside arc like a compass needle; those used on convex edges, such as the edges of round tabletops, were called sunplanes presumably because they were used on an outside arc like the sun traveling around the earth.

Until about the 1870s, or later in some areas where the shop was old and had been equipped over a period of several generations, planes were made with wooden stocks. Indeed, there are extant in certain collections a few wooden stocked planes made by carriage makers, who were close kin to the cabinetmaker, employed by the automobile companies until the 1930s. But after the ironstocked plane, with its more accurate, more durable sole and improved adjustments, was generally adopted at the end of the nineteenth century, many of the cabinetmaker's handmade planes were replaced by

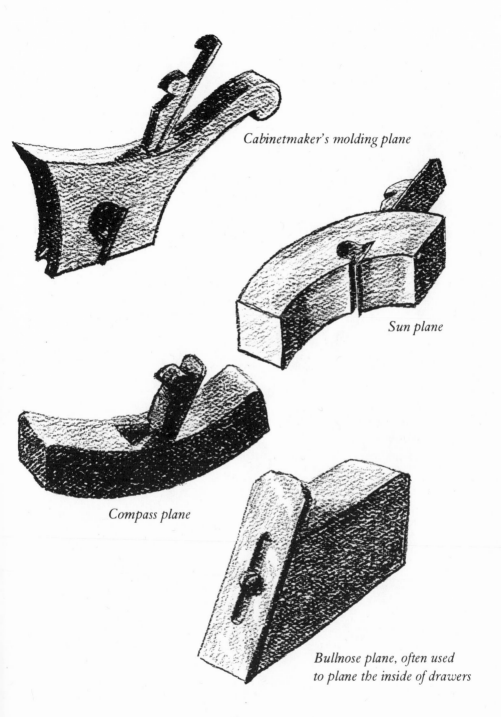

Cabinetmaker's molding plane

Sun plane

Compass plane

Bullnose plane, often used to plane the inside of drawers

Curved molding on a clock cabinet

Clock cabinet door finished with sun and compass planes

mass-produced multipurpose planes. Most molding could then be shot by one plane stock which accommodated a number of molding irons. One patent plane with a springy adjustable sole took the place of the bewildering variety of sun and compass planes used before mass production took over the toolmaker's trade and made it into a full-fledged industry.

The period of the mass-produced cabinetmaker's planes was relatively short-lived, however. The same mechanical processes and principles, the demand which could only be satisfied by mass production, imposed all manner of whirring bench tools in the furniture maker's shop and transformed it, too, into a small factory where machines were far more important economically than the artistry of the individual craftsman.

Because of the wide use of glue in joining parts of furniture, most cabinetmakers, particularly those of nineteenth-century America who made veneered furniture, had what was called a toothing plane. With a wooden stock almost identical to that of a smoothing plane, the toothing plane had a vertically wedged blade with a toothed edge. It was pushed over the surface to be glued, scratching deep grooves in the wood to serve as keys to provide more surface for the glue. Sometimes the toothing plane was used to reduce the surface of a rough board, as with the jack plane.

Toothing plane, used to prepare broad surfaces for gluing, or to roughly reduce the thickness of a board

Most cabinetmakers also owned at least one routing plane. This specialized tool was used to route out a smooth surface behind carvings, or to clean up the corners of a groove or long mortise. In appearance it seemed a combination of a plane and a spokeshave, with an angled blade wedged in the stock and a pulling handle on either side of the blade. It was also known as "the old woman's tooth."

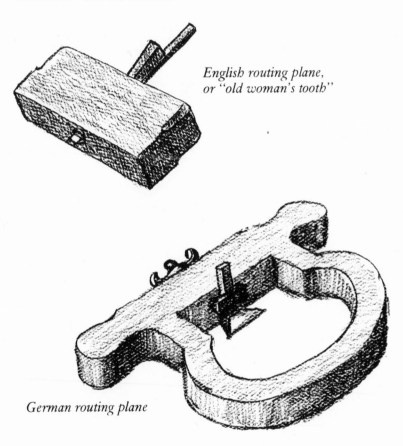

*English routing plane,
or "old woman's tooth"*

German routing plane

Carved decoration on front of an 18th-century chair on which a routing plane was used to create relief

Witchet

Witchets were not exactly planes, but functioned on the same principle as planes in rounding square or rough stuff into large dowels for plain chair legs, stair banisters, stair rails, and other such pieces. They consisted of two pieces of wood, which served as stocks, each with a V notch cut into it and each with a flat blade wedged into an aperture in the stock. Since they operated in conjunction, the parts of the witchet were fastened together by two wooden screws so that the distance between them could be adjusted. When adjusted the witchet could be placed over the end of a square board and turned to cut off the corners. It was used to cut tenons on square boards as well as to round the full length of square boards.

The late nineteenth-century form of witchet consisted of a wooden stock equipped with a handle on each side, with a hole bored through it, and a throat sawed into the side of the hole. A plane blade, held at an angle in the throat with a screw or a wedge, performed the cutting. These were used with a rotary motion to cut a tenon, each size requiring a different cutter.

For cutting round tenons only on round stuff, the late-nineteenth-century cabinetmaker also resorted to the patented metal tenon cutter that had been developed for the wheelwright, adjustable as to size and used in a brace.

Round stair rail made with a witchet

Scrapers were usually improvised as needed (then saved for future use) by shaping a bit of old saw blade to fit a molding and clamping it in a block of wood, offset in shape to form a fence. The scraper blade was secured in the block with a wood screw or bolt and nut, or even a clinched nail.

Until machines began largely to replace hand tools several auxiliary pieces of planing equipment continued to be used, particularly in the close joining of mitered corners. The miter vise was cut precisely to a 45° or other angle in which a miter was secured after sawing and planed exactly to the desired angle. Miter blocks were clamped by a beautiful wooden screw, all parts being made in the shop.

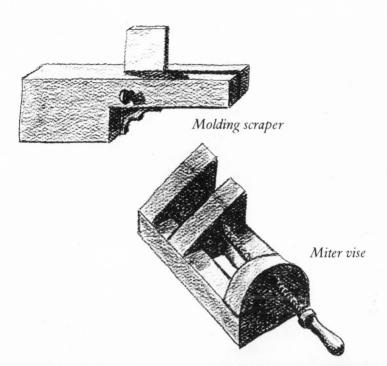

Molding scraper

Miter vise

Walnut molding finished with a scraper

Molding miter finished with a miter vise

The molding board was merely a refined version of the carpenter's shooting board, constructed for permanence out of good hardwood, carefully chosen for grain that would not warp. Such a tool was necessary for planing the small moldings that decorated furniture. A molding board could also be adapted for planing specified angles along the edges of strips intended for molding and trim. When used on a molding board the plane was held on its side, the edge of the blade then being perpendicular to the bench top.

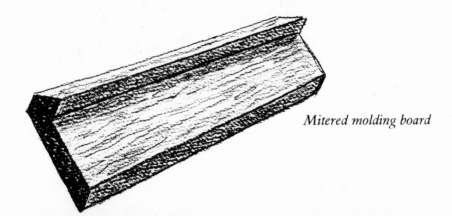

Mitered molding board

*Corner cabinet with corner boards planed
on a mitered molding board*

A late-seventeenth-century worker would have collected an astounding number of sizes and types of chisels, and every variety was retained by subsequent workers until, by the last half of the nineteenth century, a well-equipped bench might have a hundred to a hundred and fifty chisels and carving tools. Chisels are still necessary for finishing even in a well-equipped late-twentieth-century cabinet shop.

Forming and paring chisels were not one whit different from those used by the carpenter except that the cabinetmaker might have a few smaller sizes. The same comment applies to mortising chisels, for all of these were used for exactly the same functions in house building and furniture making.

Another very special type of chisel was used to hollow out the lock cavity inside a drawer. This was made entirely of steel with virtually no handle since it was double-ended. It consisted of a quarter-inch steel rod nine inches long drawn out to a chisel edge on each end, the edges being perpendicular to each other. Each was bent in opposite directions perpendicular to the axis of the rod two inches from each edge. The result was a roughly S-shaped tool, its edges pointing in opposite directions, one edge being perpendicular to the axis of the rod, the other being parallel to the axis. Such a tool could be easily placed to chisel out the edges of the lock cavity when being tapped with the Warrington hammer. It could be used on very small drawers since it took up only two inches of space above the cavity inside the drawer and could be hammered on the portion of the rod that extended outside the drawer.

Lock chisel

Lock cavity formed with a lock chisel

The chisels that separated the skilled carpenter from the hyperskilled furniture maker were the carving chisels required in infinite variety by the furniture maker for the delicate carving of chair backs, table legs, picture frames, and mantels. In this sort of work, tools are quite as important as skills and experience. Concomitantly, the work that produced the tools, the forging and tempering and grinding, was often as delicate as the carving for which the tools were intended.

As a consequence, some cabinetmakers who specialized in fine wood carving had small forges in their shops, not trusting even a skilled blacksmith to produce a carving tool that was no more, at the time of need and concept, than a vision in the carver's mind. Since they were small, sometimes a carving chisel might be conceived and executed for only one or two special cuts in a carving, the time needed to forge it being

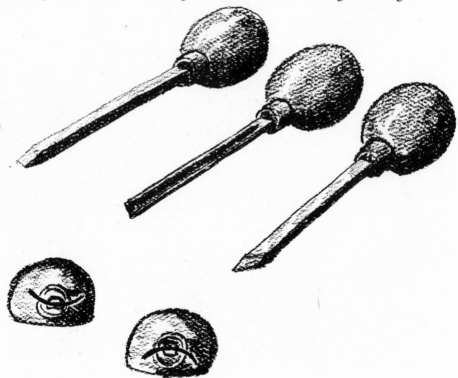

Carving chisels: Flat—Pointer—Skew—Gouge—Reverse gouge

A carved rocking-chair post

Carved decoration on a chair back

considered insignificant compared to the time needed to finish the carving with an inappropriate tool, or compared to wasted hours caused by bungling a carving in the final cuts with an ill-suited tool. The careful design and precise function of carving tools reflected as much as anything the attitudes of the hand craftsman in relating the time required to do a job properly to the lasting values of the work, which was usually judged in terms of generations.

So the wood-carvers among the cabinetmakers might amass hundreds of carving chisels during a long and satisfying career, many of them so specialized that they might have been used for only one job. The collection would have the basic small paring chisels, gouges of varying arcs, pointers and skews. But in addition there would be all manner of marvelous, gracefully shaped goosenecks, some for gouging and some for paring; curved routing chisels; special pointers; reverse gouges with the bezels ground opposite to the normal gouges. All expert carvers soon learned the importance of chisel handles and made their own from fine wood to fit only their hand, both for comfort and control. Important, too, in the proper maintenance and use of carving tools was the equipment for keeping them razor-sharp: the grindstones, whetstones, honing stones, and strops. These, since a dull tool was no tool at all, were kept and maintained as carefully as the chisels they sharpened.

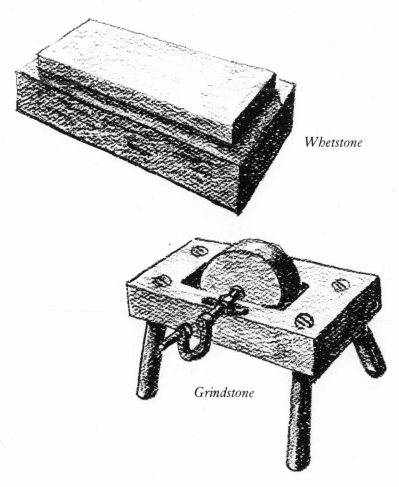

Whetstone

Grindstone

Occasionally the cabinetmaker, like the carpenter, would use the ancient auger for boring holes to receive round tenons, or to start a large mortise. The auger, however, was designed for the heavy work of housewrights, millwrights, and shipwrights and rarely suitable for the smaller work needed for furniture. There were, however, several other boring devices, all somewhat mechanical in concept, that were well established in cabinet shops by the time America was settled.

Two of the oldest of these devices were the **pump drill** and the bow drill, both operated by reciprocal rather than continuous action. The pump drill consisted of a chuck, to hold the boring bit, a flywheel of wood or stone, a shaft inserted through a horizontal movable handle, and a cord threaded through a hole in the end of the shaft and fastened to the ends of the handle. To operate, the worker twisted the thongs around the shaft, causing the handle to rise. When the bit was placed on the work the handle was brought down, untwisting the cord and rotating the shaft. The momentum of the flywheel then twisted the cord in the other direction and the handle was again brought down, rotating the shaft. Pump drills had not enough force to bore holes larger than about a quarter inch, but this was a most suitable size for the trunnels or pegs that secured tenons in furniture mortises.

For larger holes the **bow drill**, another ancient tool, was employed. This consisted of a spindle in which the bit was fastened with a loose, revolving button set on a pin in the top of the spindle. The motive power was furnished by a steel or wood bow equipped with a loose string and with a handle at one end. To use, the operator held the button on the spindle, pressing the bit downward in the wood. Then he wrapped the bowstring once around the spindle and by drawing the bow backward and forward caused the spindle and bit to rotate quickly in opposite directions until the hole was bored. Both of these devices were ubiquitous in cabinet shops until the middle of the nineteenth century, and became obsolete only in the latter years of that century. They required great skill to hold steadily and bore a straight, accurate hole.

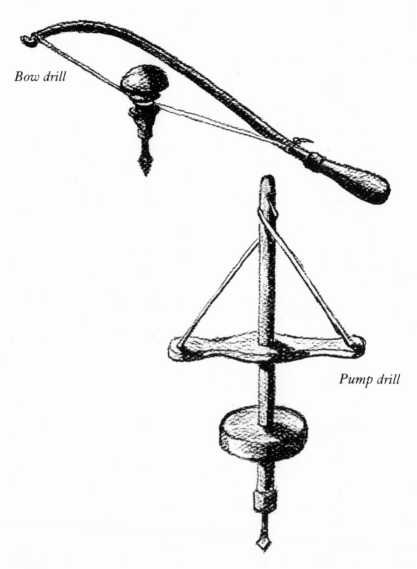

Bow drill

Pump drill

The brace, which allowed a continuous rotation in one direction, was used by the cabinetmaker more than by the carpenter. It consists of an elbow—made of wood in early times, and of metal in modern times—with a chuck on one end, a spindled button on the other, and the elbow in between. As the button and the chuck are aligned, pressing on the button and cranking the elbow in a clockwise direction enables the bit to cut or scrape a hole in the wood being worked. The brace illustrates one of the simplest of mechanical movements. It is possible that the principle it uses was an unconscious inspiration for the crankshaft now found in virtually all internal combustion engines.

The first braces were apparently no more than naturally shaped crooks in tree limbs, refined by adding the button to one end and by cutting a square hole in the other to serve as a chuck, in which the bit was secured by a pin or merely by pressure and wedging action. Toward the end of the eighteenth century, when toolmaking became an established trade, beautiful braces of fine wood were provided the cabinetmaker, the crank handles turned on a lathe, the buttons turned from the centers of logs, the chucks improved and bound with metal. Brass reinforcements were screwed or riveted at various strategic points to strengthen the stock. In brief, the brace became a work of art in itself, as finely made and durable as any piece of furniture for which it was to be used.

Later, in the early nineteenth century, the chuck was improved with a thumbscrew to secure the bit and later still with a spring catch which automatically gripped the bit shank when it was thrust into the chuck and which was released by merely pressing a button.

During the eighteenth century the carriage makers developed what was called the basket brace because its button, designed to be pressed by the chest of the worker, somewhat resembled a conical basket made of iron. Its stock was also made of iron and it probably served as the prototype of the steel brace with wooden button and free-turning handle which gained acceptance by woodworkers after 1870. An improved split chuck, tightened with a thumbscrew, was also developed about 1870, and by 1900 most braces were equipped with a convenient shell chuck with adjustable jaws. Also in the latter years of the nineteenth century the ratchet brace appeared, bringing the tool to its ultimate form until the power drill with wood bits became overwhelmingly popu-

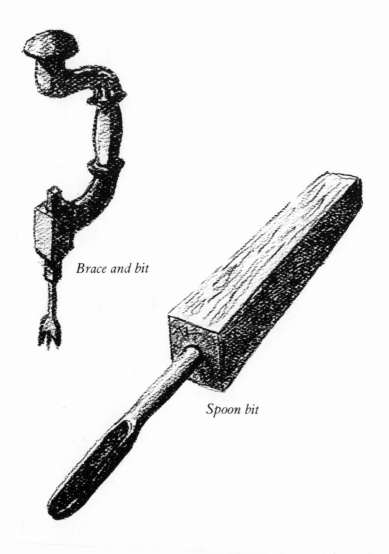

Brace and bit

Spoon bit

lar. The ratchet brace allowed the tool to be used in tight corners because the rotation could be limited to a quarter-turn or half-turn without disengaging the bit at each counter-turn of the crank.

While speaking of boring tools in the cabinet shop one might note one technique of joining the round rails, legs, and stretchers of turned chairs, where the mortise was a simple bored hole and the tenon rounded to fit it. Such joints were

Fine chair joined with round tenons

Rocking chair with round tenons

Country chair with round tenons

common among the chairs made in the backwoods of parts turned on a pole lathe. The same construction was used in making the famous Hitchcock chair in New England from the 1840s until after the Civil War.

Pieces of both the backwoods chair and the Hitchcock were secured by utilizing the natural properties of wood; no glue, nails, or pegs were needed to make joints unaffected by wear or weather. The technique was simple and logical: All members of the chair with mortises, such as legs and back, were made of green wood; all tenoned members were formed of well-seasoned wood. When the chair was assembled it was set aside for a month or two to allow the green wood to season. While seasoning it shrank, binding the dry tenons so tightly that the chair seemed to have grown together.

Of course, this technique was possible only in an area where the wood for chairs could be cut shortly before forming. It was impossible to apply on the imported woods used by fine cabinetmakers.

SCREW TAPS AND BOXES

The early American cabinetmaker used wooden screws on spinning wheels, linen presses, and certain tools used in his own shop, such as marking gauges and various clamps and vises. All these wooden screws could be made in his own shop, and until the nineteenth century many of the tools which were used to make the screws were also made by the craftsman himself. He followed some simple, logical techniques which can be traced back at least to the Middle Ages, and which may be suspected to have originated in the times of those consummately ingenious engineers, the Romans. Indeed, the concept of the screw may be traced back to ancient Greece.

Screw taps come in two forms. The first, used to cut threads on the inside of a hole no more than an inch in diameter, was made of steel and resembled an auger. It consisted of a tapered square bar with its corners filed to form pyramidal teeth which spiraled upward to a shank. The shank, with a square or rectangular end, was clinched through a stout crossbar exactly as augers are handled. A metal screw tap varied in length from six to about fifteen inches. To use it one merely inserted its end in a hole which was the same diameter as the diagonal measurement of the tap at the base of the teeth at its widest part.

Threads in larger holes, even up to six or eight inches for such equipment as cider presses, were cut with a round wooden tap through which a steel cutting tooth was wedged. The pitch of the thread was controlled by a spiral saw cut which led to the cutting tooth. A crossbar at the top and a box equipped with a small steel plate at the bottom completed the tap. To use, the workman disassembled tap and box, inserted the wooden post through a hole of the same size in the block to be threaded and again into the box. Then the steel plate was engaged in the spiral saw cut and the post was turned so that the spiral cut pulled the post downward, and the cutting tooth followed to cut an inside thread. On larger holes several cuts might be necessary, the cutting tooth being extended perhaps an eighth of an inch for each successive cut.

Metal screw tap

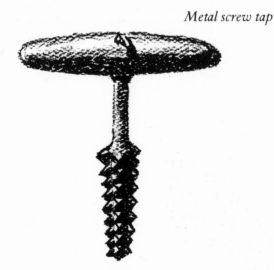

*Wooden screw tap
with steel cutting tooth*

Screw boxes were made of two rectangular hardwood boards of from one to two inches thick. Each was pierced with a hole in precise juxtaposition when the boards were screwed or·pegged together, the lower hole being of the same diameter as the dowel being threaded, the hole in the top board being already threaded. A cutting tooth of steel was secured by wedge or screw in the joint of the boards, its point projecting into the hole to the depth of the threads in the upper hole. Larger screw boxes usually were equipped with turned handles mortised into the end of the box. To use, the box was set on a dowel held in the vise and the box was turned clockwise while being pushed downward. As soon as the newly cut thread on the dowel engaged in the threads of the box, the rotation of the box guided the cutting tooth for the rest of the way.

*Screw box or die
with steel cutting tooth*

When the products of screw-cutting metal lathes became easily available to craftsmen all over the country after 1840, toolmakers began to offer metal taps of all sizes, made on a lathe and finished by filing a cutting notch through the threads. These were then used to make the screw box by the ancient method.

Since wooden screws did not lend themselves readily to mass production, they became obsolescent around 1900 and were replaced by metal screws in wood vises and clamps. A metal vise screw is undoubtedly as efficient, and stronger than the wooden screw, but it cannot easily be made by hand and certainly lacks the ingrained beauty of the old wooden screw.

Wooden screw-cutting taps and dies have continued to be available in England and on the Continent, though these tools are now mass-produced. They are offered to Americans, mainly hobbyists, through one or two distributors of hand woodworking tools.

Wooden screw and nut

DRAWING OR SHAVING TOOLS

Drawing tools, or cutting tools that function when drawn toward the worker, probably had no counterparts in the Stone Age, but they apparently have been widely used among woodworkers since man first learned to shape and temper iron and steel. Foremost among them is the **drawknife**, used by carpenters, shingle makers, wheelwrights, and cabinetmakers over the ages. It is no more than a cutting blade of various lengths, with tangs forged at each end that are bent at right angles on the same plane as the blade and then fitted with turned wooden handles. In cabinet shops the drawknife was mainly used to rough out the shapes of furniture components such as chair arms and kneed legs.

Some drawknives were designed with curved blades and were usually referred to as scorps or scoops. These were used to shape the bottoms of wooden seated chairs such as Windsors so that human occupants would sit comfortably.

Wood being worked with a drawknife was held in the vise or on the bench top with a metal holdfast. Often, however, it was worked while clamped in a shaving horse which consisted of a stout bench, often no more than a puncheon with peg legs and slanted top, equipped with a pivoted vertical clamp operated by a pedal beneath the bench. This ingenious device allowed the craftsman to sit while he worked and gave him instant clamping or release for the piece which he was forming.

For finishing irregularly shaped pieces the cabinetmaker

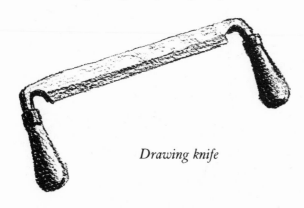

Drawing knife

resorted to the **spokeshave** in its various forms, the name of which suggests its possible origin and wide use in the wheelwright's shop. It was, however, quite valuable for the delicate shaping required in a cabinet shop and considered a precision tool because its blade, like that of a plane, could be adjusted to control the thickness of its shavings. Early spokeshaves had wooden stocks, with tanged blades, the tangs being bent perpendicular to the plane of the edge. The tangs were inserted in two holes in the stock, and the depth of the blade was controlled by jambing the tangs. Most spokeshaves made a flat cut, but some were shaped into arcs to cut a shallow

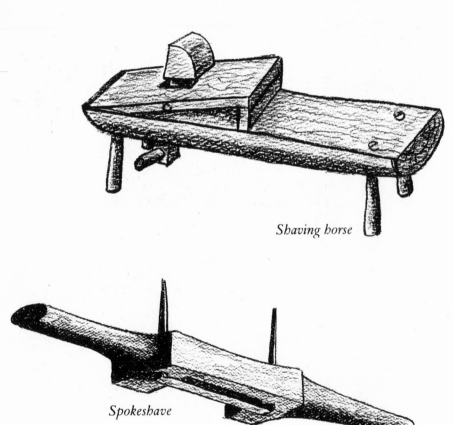

Shaving horse

Spokeshave

Chair leg formed with a spokeshave before carving

Chippendale chair back shaped with a spokeshave

groove or to shape a concave shoulder. Late in the nineteenth century the toolmakers developed spokeshaves in which the depth of the cut was controlled by two nuts which fitted on threaded tangs. Still later, at about the time molded iron stocks were adopted for planes, spokeshaves were made with iron stocks and with a double iron, as in modern planes, to break up the shavings as they were separated from the stuff being worked.

Scorpers were made on the same principle as the spokeshave except that they had but one handle and, of course, a convex blade to make a concave cut. They were ideal for finishing the shape of Windsor chair bottoms.

Scorpers had almost entirely disappeared from the cabinetmaker's toolbox by about 1900. Spokeshaves may be considered obsolescent in the latter half of the twentieth century, although a few are still being made and sold to the few workers and hobbyists who can identify them and understand their function. Only the drawknife, still in its pristine form, is read-

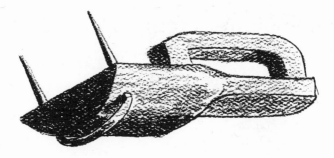

Scorper

ily available in most modern hardware stores. But even its future is uncertain in a day of machines with rotary cutters that can be adapted to almost any job of shaping wood.

TURNING LATHES

The most interesting of the mechanical devices found in early American cabinet shops, a machine that dated back at least to the Middle Ages and perhaps beyond, was the lathe, powered by foot or by a great wheel or by water. It was used to turn round chair and table parts, to turn great bedsteads, or to shape dishes and standards for wooden candlesticks. Its operator was known as a turner.

Undoubtedly the earliest form of this machine was the **pole lathe**, which remained in use in Europe and in certain isolated areas of the American Appalachian region until World War II. It consisted of a simple stock made from two logs or boards inserted in a dirt floor or equipped with standards for a wood floor. Fastened to the ceiling above was a springy pole or lath (which gave the machine its name) to the end of which was fastened a piece of rope or a length of leather strap. This was wrapped once around the stuff to be turned

Pole lathe

and fastened to a simple treadle which had one of its ends secured so that it pivoted. A stroke of the turner's foot rotated the stuff so that it could be cut with a long-handled turning chisel; when the treadle was released the spring pole caused the stuff to be rotated in the other direction, the chisel being withdrawn until the next downstroke on the treadle. Its principle of operation was exactly that of the bow drill and pump drill.

*Chair post capital
turned on a pole lathe*

Effective, though not efficient, the pole lathe was retained in most cabinet shops until about 1800. It was supplemented, however, around 1700 by the **treadle lathe** equipped with flywheel and belt, which allowed continuous turning in one direction. Both the pole lathe and the treadle lathe are most satisfying machines to operate, for the regular rhythm of the turner's foot action is soothing and satisfying and seems to inspire the artistic rhythm of the pieces he designs as he turns.

Most of the fine American cabinet shops of the eighteenth century were equipped with the **great wheel lathe**, with a stock no different from other lathes, but with a detached flywheel of perhaps six feet in diameter, set in its own stand several feet from the headstock of the lathe. The great wheel was operated by an apprentice who was told by the turner when to speed up the action or slow it down to suit the character of the wood being worked.

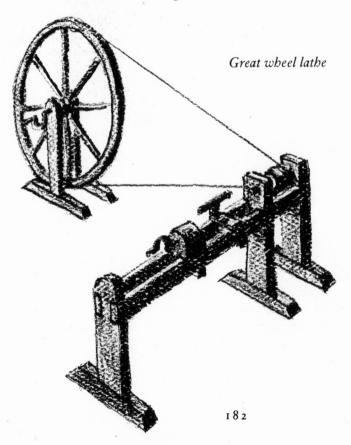

Great wheel lathe

Treadle lathe

Of course, waterpower had been available to operate machines since Roman times, and whenever a cabinetmaker could locate his shop near a stream he saved his energy and that of his apprentices by using waterpower to turn his lathes. Substituting one source of power for another was an easy matter, requiring no change of attitude when steam and electricity and internal combustion appeared in the vestigial factories of the nineteenth century. The lathe form has been retained with no startling changes in design or adaptation.

Bed turned on a great wheel lathe

TURNING CHISELS

Like the woodcarver, the turner used a variety of chisels in his work. Also like the woodcarver, he often made his own, carefully forging, grinding, and tempering them to suit his taste, making his own handles on the lathe to fit exactly the size and contours of his hand. With such custom-made tools he was able to impart a certain indefinable delicacy to his work that would have been impossible to create with factory-made tools; he was able to incorporate his own personality and artistic aspirations in each turning he made.

These turning chisels were found in awesome variety. Some were pointers, some were flat-edged, some had edges ground askew. There were gouges in every size, some for roughing, like the iron of a jack plane, and others to provide just the right-sized, right-shaped indentation to emphasize dimension or create proportion in a turned bedpost. Usually the blades were short and tanged and inserted into a rather long handle, long enough at least to be grasped by two hands for utmost control over the tool. The chisels were usually kept on a shelf or rack over the lathe so that they could be exchanged quickly as the work progressed, before the bright fires of momentary inspiration grew dim.

Turning tools

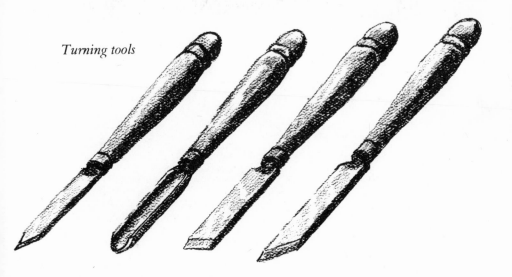

Whatnot of pieces turned on a treadle lathe using a number of turning chisels

FASTENING

Seldom did the maker of fine furniture resort to nails, or even to pegs for basic fastening. He used an animal glue, refined from horns, hoofs, and hides into a smelly adhesive kept in a glue pot which was heated when the glue was applied to mortises and tenons, dovetails, veneer surfaces, and small pieces of molding. Small molding brads were frequently used to keep molding in position while gluing, and pegs were used on heavy mortises and tenons to tighten the joint, but the strength of most joints depended on a precise fitting and a spot of hot animal glue. Hot glue was used because it set up very quickly.

Unlike the miracle epoxy adhesives which appeared during the 1950s, the effectiveness of animal glue depended on tightly clamping the pieces being joined until the glue dried, usually after setting twenty-four to forty-eight hours. This was done with a variety of clamps of different shapes for different purposes, most of them being made of wood, but a few being made of iron.

For gluing boards together at their edges to create a wood surface large enough for a tabletop or desk leaf, the **bar clamp** was used. This was a wooden bar with a series of holes or notches along its length into which a peg or rest could be inserted, and a headstock, through which a wooden screw was inserted, mortised into one end. The adjustable peg was inserted into a hole which nearly matched the width of boards being joined in its distance from the headstock. After the boards had glue applied they were laid flat on the bar and the screw in the headstock turned to clamp them tightly together until the glue set. Bar clamps were also used to clamp the rails and legs of chairs and tables.

Iron bar clamps most often consisted of a six-foot-long bar or rod, three-quarters of an inch square, with one end upset to form a button and then bent back so that the button pointed down the length of the bar. A rectangular iron plate, pierced at one end with a hole which allowed it to slide up and down the bar, and equipped with a hand screw on the

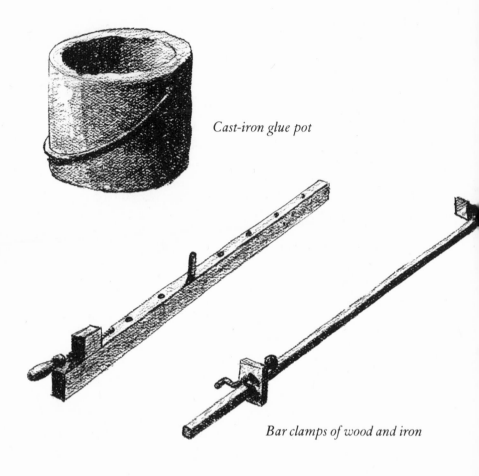

Cast-iron glue pot

Bar clamps of wood and iron

other end, was placed at the proper position, the work placed in the clamp, and the screw tightened. Tightening of the screw jambed the plate in position, holding the joint securely until the glue had set.

For smaller pieces, a clamp of two stout, short boards of square sections, each held to the other with two wooden screws, was used. The screws pointed in opposite directions so that both hands could be used to adjust and tighten them. Often several of these clamps might be used to press a contoured board on top of the veneer being glued to the front of a bowed or serpentine drawer front.

Still another type of clamp was no more than two flat boards of from two to three feet long, held together at the center with one wooden screw. Such flat clamps could be used in gluing a long piece of molding to the bottom rail of a chest of drawers or for similar jobs.

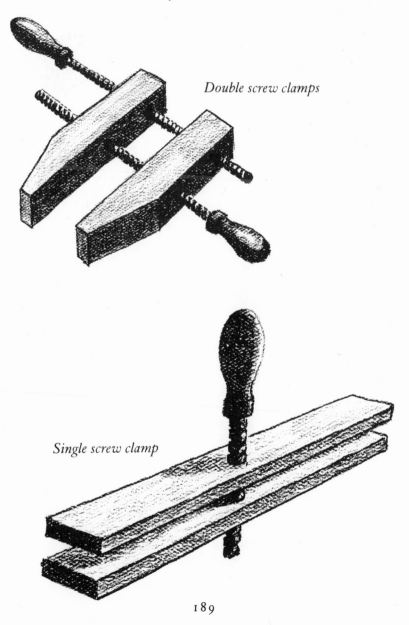

Double screw clamps

Single screw clamp

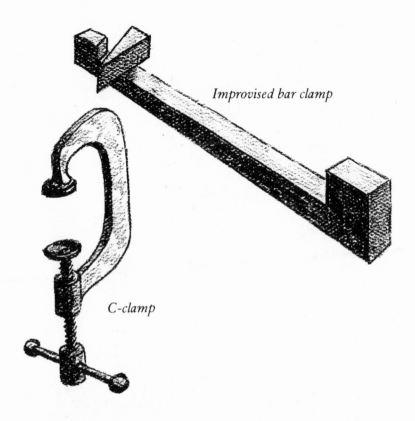

Improvised bar clamp

C-clamp

From about 1800 until the present the cabinetmaker frequently relied on the C clamp, probably the most familiar of these devices to twentieth-century hobbyists. It was not nearly so common before 1900, however, because it had to be made in a blacksmith shop. The old cabinetmaker preferred making his own clamps, and he also preferred using wood on wood.

From time to time even the finest craftsman sometimes finds himself without the exact tool needed for a certain job and as a consequence he improvises. This was particularly true for clamping. When necessary the cabinetmaker resorted to several types of clamps which could be made quickly, their basic feature being the substitution of the primitive wedge for the screw. These improvisations were somewhat awkward, but nevertheless efficient, because the same principle was applied; after all, a screw is actually no more than a wedge wrapped around a post for convenience and latitude.

A Chippendale chair with glued joints

Sometimes the cabinetmaker would nail two almost parallel boards to the bench top, place the pieces being glued together between them, and wedge them tightly together. At other times he would take a long board, cut a shallow notch in most of its length slightly longer than the width of the pieces, and tap a wedge into the leftover space. Small clamps of the same basic design could easily be made for gluing molding or other small refinements to a piece of furniture.

MISCELLANEOUS TOOLS

All the woodworkers used marking tools such as scribers, marking gauges and cutting gauges, squares, *et al.*, which have already been described as part of the carpenter's tool kit. In addition to these the cabinetmaker used a couple of other tools which fall into the miscellaneous category, calipers and the bevel. Those who also upholstered furniture made frequent use of an upholstery stretcher.

Mortising gauges were equipped with two staves, each with its own scriber point set at different distances from the block so that the two points together could accurately mark the width of all mortises on a piece of furniture. Cutting gauges were of the same basic design but equipped with a small knife point instead of a scriber. They were used to make shallow cuts of mortises, or to cut a precise line in veneer.

Mortising gauge

CALIPERS

Several pairs of calipers were needed in every cabinet shop from early Colonial times until at least World War I. Of course, the turner, working at his banister or other turnings in manifold quantity, required not only straight calipers but also those with bent points for measuring external and internal diameters. Before the beginning of the industrial age, calipers were made by the blacksmith to the specifications of the furniture maker, or of scrap wood by the cabinetmaker himself. Often in this imaginative period they were decorated as beautifully as a piece of table silver, being carved with flowers or scrolls or grotesquerie.

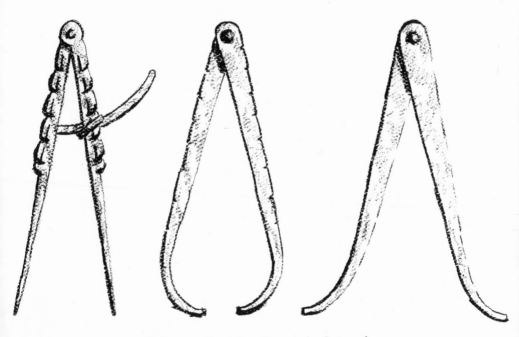

Calipers—Straight—External—Internal

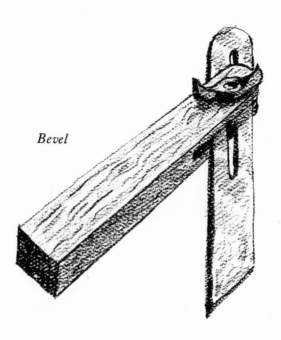

Bevel

BEVELS

Bevels consisted of a short, relatively thick board, split or mortised at one end with a second pivoting board or strip of metal which was inserted into the split and fastened with a thumbscrew. The thinner arm of the bevel could then be set at any desired angle and secured. It was then used like a square and the set angle marked with a scriber for sawing or chiseling.

UPHOLSTERY STRETCHER

Most American furniture makers in early times also furnished the upholstery and had around the shop that peculiar tool of the upholstery specialist, the cloth stretcher. Resembling a large pair of pliers with wide jaws and a spur under one of them, this tool was used by grasping the edge of upholstery cloth or leather between the jaws and using the spur as the fulcrum of a lever to stretch the material tightly before tacking it to the wood. This tool is one of the few which has not been replaced by a machine.

But most of the others have been replaced in ways that an eighteenth-century cabinetmaker would have thought impossible.

A few have changed only through the substitution of electricity, steam, or internal combustion for foot and hand power. Hammers and mallets are essentially the same as those used in the Stone Age. Axes and adzes, however, have been almost totally replaced by power saws, even in the home workshop of the amateur cabinetmaker. Bow drills and pump drills are obsolete and the electric power drill has made the brace obsolescent. Even the plane, that marvel of engineering ingenuity, in all its graceful, admirable forms, has been displaced to a major extent by whining, dangerous machines which carve the same molding in one-tenth the time, and with one-tenth the degree of satisfaction to the worker. The basic form of the lathe has remained unchanged, but it has been adapted to the needs of mass production, powered by electricity, tended in large modern furniture factories by a computer instead of the precise and patient turner who used to go home to his family each night with varicolored chips in the folds of his clothing, smelling deliciously of oak and walnut and pearwood and pine.

Upholstery stretcher

The wood-carver, artistically creating his floral designs on the knees of chair legs and on the backs of chairs and the heads of beds, is generally employed in modern times only to make prototypes, and these must be designed so that they can be duplicated by machine. Machines do not use chisels of infinite shapes; carving machines are equipped with rotary points which trace the original work and duplicate it. Such mechanization cannot duplicate the quality of fine carving, but it carves many pieces in one-tenth the time it takes to make the original and, importantly, at one-tenth the cost.

Most of all, and sadly so, the cabinetmaker's shop has been generally displaced by the factory. The shops in America where furniture is made entirely by hand, piece shaped to fit another hand-formed piece, are rare indeed, and are for the most part relics found only in restorations such as Williamsburg, Old Salem, and Westville. There are indeed cabinetmakers listed in every large telephone book found today, but most of them make only kitchen cabinets, and those who produce the rich walnut paneling and bookcases for law and business offices are really no more than miniature factories equipped with the most modern of power tools. No more does one find the shop floor carpeted with rich brown, curly shavings so useful in starting the fire on a wintry morning or for keeping the glue pot warm. The floors of these shops are strewn with the fine, useless chips from rotary blades.

But how could it be otherwise? Where could one find in modern America the required number of men capable of being trained for such exacting work, and how could even an army of such men produce enough tables and chairs, beds and sofas, desks and bookcases to furnish all the new houses and apartments that spring up so quickly in today's cities and suburban areas?

The furniture industry has changed from hand to machine because of unavoidable circumstances. The machine age has been an inevitable and inexorable consequence of the eighteenth-century scientists and inventors. Science, through its economic effects, has created new needs which in turn have created new materials and industries and technology.

18th-century upholstered stool

Take, for example, the decline of the classical mortise and tenon joint. It had to be made by hand, which took time. In the early nineteenth century, when large factories first began to appear in America, the dowel joint began to be used instead of the mortise and tenon. Holes for the dowels could be drilled easily and quickly by a drill press equipped with simple jigs. With the new technology dowels could be mass-produced to fit the holes precisely. A joint almost as good as the mortise and tenon can be made without the use of time-consuming handwork. And so the mortising saw and mortising chisel began to disappear. In the late twentieth century the mortise and tenon joint has reappeared in better factory-made furniture, but only because machines have been devised which can cut them precisely with a minimum of hand labor. In essence the change has been no more revolutionary than the development of the plane long centuries ago to do more quickly, and really better, the jobs for which chisel and drawknife had been used for long centuries before. Everything is relative.

New materials, the products of new needs and expanding science, have also served partially but importantly as the nemesis of handcrafted furniture. Much of the furniture of the late twentieth century has completely eschewed the use of wood, that basic God-made material which once was the exclusive material for the furniture maker. Much modern furniture, particularly that used for porches and gardens, is made entirely of steel or aluminum or magnesium, or a combination of these metals with plastic fabrics. Stamped sheet metal is used for whole cabinets, tables, and wall shelves, with drawers and decorations formed on huge presses and assembled with blue-flashing electric arc welding machines, then sometimes finished with wood-grained decalcomanias. These pieces, gauche though they may be, do serve a need and provide inexpensive furniture for those who otherwise might not have any furniture.

Expensive living room furniture in fashionable New York apartments and contemporary showplace homes are often made entirely of formed plastic on intricately shaped, laminated wood. Plastic bags filled with water have to some extent displaced the beautiful turned bedsteads of Mount Vernon and Newport.

Indeed, plastics, an industry which saw its obscure beginnings in false ivory for billiard balls in the late nineteenth century, has had great effect on the style and methods of manufacturing furniture, particularly in the kitchen and that child of running water, the bath. Laminated phenolic, such as Formica, which is heatproof, waterproof, and shockproof, now rules almost supreme for use on kitchen countertops, bathroom tabletops and breakfast tabletops. New plastics are continually being developed today and many of them are being increasingly used as components in furniture, some of it very fine and expensive furniture.

There was a time in the eighteenth and nineteenth centuries when toolmaking firms issued catalogs displaying gentlemen's saws and tool kits, reflecting the interest among the gentry in relieving the pressures of business with handicraft. Several crowned heads of eighteenth-century Europe were ardent craftsmen, notably King Fredrik-Wilhelm III of Prussia, who kept a small wood lathe in his bedchamber, and King Adolphus Frederick of Sweden, a somewhat lackadaisical ruler but an enthusiastic cabinetmaker with a well-equipped shop. Cabinetmaking has also been a favored hobby of Americans throughout the centuries, and was a necessity for the backwoods farmers of the American frontier before the railroads and Sears, Roebuck made it possible for them to furnish their homes with factory-made pieces. But the hobbyist, too, has succumbed largely to the ease and rapid production of the machine. The electric drill has taken the place of the once ubiquitous brace and bit. Electric-powered saber saws do the work of the coping saw, keyhole saw, and foot-powered jigsaw. All-purpose electric tools, which can be adjusted to special needs by turning a handle and flicking a switch, allow one cleverly designed unit to serve as drill press, lathe, plane, and saw. Such machines have great appeal to the amateur market despite the fact that a few more fingers are lost each year than would be the case with hand tools. And new hand tools, not so precise but easier to use on aluminum, plastic, and plywood, have replaced the traditional tools of the wood-working classicists. One of the most popular is the file and rasp fitted with a plane handle and offering that advantage so closely associated with modern culture, expendability; it need never be sharpened and it may be inexpensively and conveniently replaced at the nearest hardware store when it wears out.

Wood has remained the most desirable of materials for fine furniture and doubtless will maintain its position of eminence for centuries to come. The handcraftsman in wood, however, will probably never again be seen as an integral, important member of every community. Not that the fine craftsman is unappreciated in modern America. The popularity of well-made antique furniture, from farms or city mansions, dem-onstrates the lasting admiration of Americans for good handcraftsmanship. It's just that circumstances have cast the crafstman out, leaving behind only his high standards of workmanship and design.

CHAPTER

4

OTHER
WOODWORKERS

In Colonial America there were other craftsmen in wood besides axmen, carpenters, and cabinetmakers. These craftsmen unfortunately disappeared faster than the craftsmen in other trades because the products of their hands found no place in the America of post-World War I. So forgotten are they that their tools for the most part cannot easily be identified, thus taking from them a most noble monument.

At one time every village of size had a wainwright or wheelwright or cartwright who worked with the blacksmith to make and repair the wagons and carts needed to haul farm produce to the markets in town, or to transport raw materials hundreds of miles over rough wilderness tracks. Many of the tools used by these craftsmen were the same as those used by the carpenter and furniture maker, and some, such as the spokeshave, were developed in the wainwright's shop and adopted by the cabinetmaker. These venerable, ancient trades are gone because there is no place for wagons on superhighways. Only family names such as Wainwright, Cartwright, and Wright attest to the establishments of these craftsmen in bygone days.

Holding the same relationship to the wainwright as that of the cabinetmaker to the carpenter was the coachmaker. Frequently the work he was called upon to do was more delicate and arduous than that of the cabinetmaker, consisting of large surfaces made concave and convex and carefully joined with special tools to ensure grace and durability to gentlemen's conveyances, which were subjected to the same rough roads as the Conestoga wagon. The coachmaker used numbers of woodworking tools, especially planes, peculiar to his trade and art. He and his tools, strangely enough, did not disappear entirely until a generation after wagons became rare, for he was needed to make and assemble the wooden framework and refinements of automobiles before wood construction was made obsolete in that industry in the late 1930s.

Once, and until well into the twentieth century, the cooper was an important craftsman who supplied homes with noggins and piggins and provided barrels and hogsheads for industry. He, too, has virtually disappeared, his remarkable but often unrecognized skills found unsuitable to the mass production of galvanized pails, steel or cardboard drums, and all the other containers for the industrial products of modern America. The cooper's tools were as specialized as his training and evolved over the centuries to suit his needs. A few wooden barrels are still made, but not by the cooper, at least not in America. They are now factory produced, sturdy and traditionally shaped, but lacking the indefinable aura of fine craftsmanship. The new remaining coopers employed by distilleries and vineyards are used only to repair barrels.

Another major field of the woodworker was shipbuilding. A few wooden boats and ships are still being made, but full-time shipwrights constitute more of a curiosity, a relic, than a recognized trade. This has been the case since World War II when the need for wooden minesweepers, which defied magnetic mines, recalled the older shipwrights of New England and Canada from retirement. Since World War I, however, almost all ships have been made of steel, welded together with electricity and acetylene gas. A shipwright's adz is hardly the tool for constructing 600-foot tankers. Boatwrights are still to be found in some quantity, but plastic, fiber glass, and plywood have replaced the oaken and cedar strakes and ribs that were still common only a generation ago, and the modern boatwright hardly considers himself a woodworker. His tools are machine tools and his maintenance work is done largely on factory-made boats of fiber glass and aluminum.

Fast disappearing, too, are the basketmaker, the bellows-maker, the gunstocker, and the rest of the woodworking fraternity once so essential to civilized life. All of this work can now be done by machine, and the craftsman adapts his techniques to machine technology rather than developing special tools to allow his hands to do the work.

When one looks at these woodworkers' forgotten tools, one is filled with admiration for mankind. Thank heaven there are museums and restoration villages to preserve the tools and demonstrate their uses! We all know that machines do work more quickly and cheaply but we should never forget that man did without machines for centuries and can *still* do without them—if he ever has to.

BIBLIOGRAPHY

BENJAMIN, ASHER. *The American Builder's Companion.* (Reprint of the 1827 ed.) New York: Dover Publications, Inc., 1969.

BEALER, ALEX W. *Old Ways of Working Wood.* Barre, Massachusetts: Barre Publishers, 1972.

The Cabinetmaker in Eighteenth Century Williamsburg. (Williamsburg Craft Series.) Williamsburg, Virginia: 1963.

CHIPPENDALE, THOMAS. *The Gentleman & Cabinet Maker's Director.* (Reprint of the 3d ed.) New York: Dover Publications, Inc., 1966.

Bibliography

Early American Industries Association. *Chronicle.* Williamsburg, Virginia: 1933.

EDWARDS, RALPH, and JOURDAIN, MARGARET. *Georgian Cabinet-makers.* (3d ed.) London: Country Life, Ltd., 1955.

GILLESPIE, CHARLES COULSTON, ed. *A Diderot Pictorial Encyclopedia of Trades and Industry.* New York: Dover Publications, Inc., 1959.

GOODMAN, W. L. *The History of Woodworking Tools.* London: G. Bell and Sons, Ltd., 1964.

KATZ, LASZLO. *The Art of Woodworking and Furniture Appreciation.* New York: PFC Woodworking, Inc., 1971.

LA FEVER, MINARD. *The Modern Builder's Guide.* (A reprint of the 1st ed. with three additional plates from the 3d ed. and with a new introduction by Jacob Landy.) New York: Dover Publications, Inc., 1969.

MERCER, HENRY C. *Ancient Carpenter's Tools.* Doylestown, Pennsylvania: The Bucks County Historical Society, 1951.

MOXON, JOSEPH. *Mechanick Exercises.* (3d ed.) London: 1703.

NUTTING, WALLACE. *Furniture Treasury.* New York: The Macmillan Company, 1924.

INDEX

Italic numbers refer to pages with illustrations.

Index

Index